WHY WE STRUGGLE TO GO GREEN

WHY WE STRUGGLE TO GO GREEN

HARD TRUTHS ABOUT THE CLEAN ENERGY TRANSITION

THOMAS MANUEL ORTIZ

Published by

Stoney Creek Publishing Group

StoneyCreekPublishing.com

ISBN: 978-1-965766-28-6
ISBN (ebook): 978-1-965766-29-3
Library of Congress Control Number: 2025910040

Cover design by Market Your Industry

Printed in the United States

CONTENTS

A VERY BRIEF GUIDE TO ENERGY TERMINOLOGY

This book was written by an American engineer, but I do not expect you, the reader, to have studied engineering. As you probably know, we in the United States stubbornly refuse to adopt the metric system in most cases. You will encounter a lot of what engineers call "units of measure" in this book. Virtually all of these will be what are commonly referred to as "metric" or "SI" units. The SI system of units is almost universally used in technical publications, international meetings, and government reports. In case you are not accustomed to seeing them, the following lists of units and metric prefixes will help you better understand the material that follows. The combination of a base unit and its prefix signifies its order of magnitude. For example, one thousand meters equals one kilometer. Written in symbols, 1000 m = 1 km. The "k" prefix signifies one thousand.

One other nuance you may not immediately recognize is the difference between power and energy. Engineers define power as energy transferred per unit time. The base SI unit of energy is the Joule (J), and the base unit of power is the Watt (W), which is equal to 1 J/s. However, the electricity industry does not measure energy in Joules, but rather in kilowatt-hours (kWh), megawatt-hours (MWh), gigawatt-hours (GWh), or even terawatt-hours (TWh). In other words, they multiply the instantaneous power being generated by an electric plant by the number of hours that power is used. You will, nevertheless, see both Joules and Watt-hours in this book.

Common Metric Prefixes
pico p 10^{-12} (one trillionth)
nano n 10^{-9} (one billionth)
micro μ 10^{-6} (one millionth)
milli m 10^{-3} (one thousandth)
kilo k 10^{3} (one thousand)
mega M 10^{6} (one million)
giga G 10^{9} (one billion)
tera T 10^{12} (one trillion)
peta P 10^{15} (one quadrillion)
exa E 10^{18} (one quintillion)

Common Energy Related Units
acre-ft (acre-foot) volume
bbl (barrel) volume
C (Celsius) temperature
F (Fahrenheit) temperature
ft (foot) length
g (gram) mass
gal (gallon) volume
h (hour) time
J (Joule) energy
lb (pound) mass
m (meter) length
s (second) time
t (tonne or metric ton) mass
W (Watt) power
Wh (Watt-hour) energy

Common Unit Conversion Factors
1 acre-ft = 325,851 gal
1 bbl = 42 gal
1 kg = 2.2 lb

1 kWh = 3.6 MJ
1 m = 3.3 ft
1 m3 = 264 gal
1 t = 1000 kg
F = 9/5 x C + 32

INTRODUCTION

Half a century is a long time, and time has a way of cultivating irony. When I was born in 1973, people waited hours in lines stretching for blocks just to buy a few gallons of expensive gasoline (Gershon, 2021). The energy panics of the 1970s marked the end of an unprecedented economic boom known as *les trente glorieuses*, the thirty glorious years of prosperity following the end of World War II. These crises were also instrumental in spurring the innovations in oil and gas exploration that eventually led to the US shale era (Wang and Krupnick, 2013). More than fifty years later, the world is struggling to rid itself of carbon-emitting fuels and governments are paying people to buy electric cars (Doll, 2023a). It is both a story of collective amnesia and a reminder of the pain involved in making radical adjustments to cherished consumption habits. These habits have been amplified and reinforced over generations as energy has become less expensive and more widely available. We scarcely think about the immense amounts of energy needed to make and do things, with one very important exception: the "price at the pump." Gasoline prices are the average American's daily benchmark of economic health, closer and more immediate than the gyrations of the stock market or the harder-to-understand effects of bond yields and interest rates. Almost everybody buys gas: they always need more *of* it, and they hate to pay more *for* it.

Nowadays we cannot help but hear constantly about the "energy transition," a supposed cure for global climate change and the accelerating depletion of existing energy resources that has accompanied our rising living standards. This book is not about how we should design the energy transition. The broad strokes of how to accomplish a shift away from fossil fuels have been painted, and they have been described in detail from a variety of perspectives (Nemetz, 2022; Hoffmann, 2012; Valero et al., 2021; Harvey and Gillis, 2022; Fiorino, 2022; Friesen, 2020; Helm, 2020; Harvey, 2018; Heinberg and Fridley, 2016). This book is written from the viewpoint that human induced global climate change is a genuine and serious concern: I accept and treat this as a given. We are quite able to measure temperatures, sea levels, ocean acidity, and other markers of climate to a satisfactory, if imperfect, level of accuracy. The fact that those measurements are changing is as obvious as the rising and setting of the sun. Coral reefs and the shells of marine animals are being dissolved before our eyes in a fizz of carbonic acid that now accumulates faster than the natural carbon cycle can restore its equilibrium concentration. Warmer, more energetic air holds more water and fuels more powerful storms. Regional rainfall patterns and crop growing seasons are shifting. Both floods and droughts, in a vexing cycle of inadequate drainage followed by inadequate aquifer recharge, are more intense and damaging. The ultimate effects of those observations are not entirely predictable, but they are evident and worrisome.

It is also undeniable that Earth's climate has always been changing, following various overlapping geological and astronomical cycles lasting thousands to millions of years. The climatic changes we are making today are layered upon the effects of these natural variations, the latter which, among other things, have influenced the appearance of ice ages. That does not, however, alter the fact that the most efficient and reliable lever we possess with which to counteract any unwelcome effects of global warming over the next few centuries is through the abatement of our own manufactured greenhouse gas emissions. The buildup of carbon dioxide in the atmosphere that was once largely the result of sporadic volcanic activity has since been dwarfed by the incessant pumping of waste gas from human industry.

Imagine the natural carbon cycle as a bathroom sink in which the water has always been allowed to flow just enough so that it is exactly balanced by the drain: the water level in the sink neither rises nor falls. One could theoretically leave such a sink unattended for years and never find a puddle on the floor. But then someone comes along and opens the faucet all the way. Suddenly, the clock is ticking down the minutes until the sink overflows. This is what humanity has done: we have opened the carbon faucet to full blast and the oceans, plants, and minerals of Earth cannot recycle it fast enough. Moreover, animals and plants which evolved over eons to tolerate relatively gentle variations in temperature, rainfall and acidity now face living conditions that have changed within mere decades in ways that will be fatal for many of them within a few generations.

A now common plea from people concerned about climate change is, "Save the planet." The planet does not need saving—we do. The Earth will be circling the Sun long after we are all gone. But everything about our civilization has been constructed under the assumption that the global climate is stable, which has been a good assumption for thousands of years thanks in part to a very specific and fortunate combination of eccentric orbital behaviors (NASA, 2020a). Our ancient ancestors may well have noticed this onset of climatic stability as evidenced by the emergence of sedentary agriculture (which was not theretofore possible) and the subsequent decline of hunter/gatherer civilizations at approximately the same time (Brooke, 2014). Economist John Gowdy even predicts that runaway climate change could consign our descendants to hunting and gathering once again (Gowdy, 2020)! If we want to give ourselves (and our fellow creatures) the best possible chance of evolving to live in a less stable future environment, we should do everything in our power now to slow the process of climatic change and learn to live within our ecological means. And we must do so before we exhaust all our easy to access energy sources, because technology—at its most fundamental level—is nothing but the expenditure of available energy to transform matter into more convenient forms. Sustainable living means less energy use and less material transformation.

I will therefore focus in this book on core aspects of the current consensus view on what kinds of lower carbon energy technologies might be deployed over the next several decades to help us turn the faucet back to a slow drip. I will concentrate on what to expect in terms of the inconveniences and tradeoffs needed to implement a transition to using those technologies as replacements for business as usual. My motivation for doing so is a combination of two realities. First, as is always the case with emerging products or markets, the people designing solutions are filled with enthusiasm, optimism, and (usually) greed that creates intellectual and ethical blind spots about their potential. Second, policymakers are never rewarded for giving people bad news, and the looming impacts of climate change contain a lot of bad news. Unfortunately, the available means to *mitigate* those impacts are also burdened by their own fair share of bad news. The public deserves a dose of brutal honesty, written in plain English, about what lies ahead in terms of our energy and economic future. It's not all darkness and dread, but if we ever decide to become serious about limiting the rate of global warming, then life as we know it is going to have to change—and not uniformly for the better.

Will we remain below the maximum 2°C target (equivalent to 3.6°F) for global average temperature increase above preindustrial levels, a target set by the Intergovernmental Panel on Climate Change to avoid the worst consequences of climate change? Absolutely not. The world has already experienced its first full year of breaching the much more desirable 1.5°C increase threshold (Poynting, 2024), absolute atmospheric carbon levels are continuing to increase, and governments are, as always, preoccupied with the other usual distractions: wars; trade disputes; sports scandals; and cultural disagreements. Recent trends toward the election of conservative governments overtly hostile to the very idea of human induced climate change, growing diplomatic, military, and economic tensions among the major world powers, and the (understandable) intolerance by consumers of inflation or economic stagnation will place severe constraints on our ability to pivot away from the fossil energy status quo anytime soon.

Does this mean that we should all simply throw up our hands and ignore the problem? Absolutely not. The more people see and feel the physical, tangible effects of a changing climate, the easier it will be to convince them to accept unpleasant lifestyle adjustments and collective sacrifice. Nothing galvanizes people into action like a bona fide emergency. It is too late for us to prevent substantial melting of the polar ice caps, sea level rise, intensification of storms and heat waves, or marginalization of large tracts of existing farmland. What we can, and must do, however, is to prepare as best we can for the consequences of these certainties and to, belatedly, slow our rate of greenhouse gas production enough so that people (and animals) will still be able to live here a few centuries from now. Alternatively, we can expend our effort, energy, and capital on a radical escape plan. A few years ago, I would have dismissed such notions as movie plots, but the fact is that influential people are beginning to spend serious money on schemes to design lunar bases, Martian colonies, and asteroid mining processes. Let's assume, just for the sake of argument, that emigration from Earth is a feasible option. If this appeals to you, please be aware that not all eight (or ten) billion of us are going to be booked on that flight to Mars when it blasts off. And, while getting yourself "off world" might be a great plan if you live in the fictional universe of *Blade Runner*, it would be a most unwelcome fate for those of us who would have to rely solely on space technology available outside of Hollywood.

I personally would rather make my stand here on Earth. For those of you who agree, what should we do to prepare? What kinds of sacrifices are people living in wealthy countries willing to accept in exchange for making the transition to a sustainable future? I suspect that it will depend entirely on what their next best option turns out to be. That next best option will depend on how long we wait to focus on climate adaptation measures and energy policies designed specifically to reduce and remove carbon emissions, rather than those that are primarily intended to stimulate economic growth. Reducing our dependence on energy-dense fuels and other scarce resources in exchange for lower quality, less reliable, more expensive sources of power and fewer consumer goods is not a winning strategy if viewed through the lens of modern market economics. If it were, we would already have done it.

The longer it takes us to accept the responsibility of climate-centric economic management, the more expensive damage will accrue to our communities and landscapes. The January 2025 wildfires in Los

Angeles are a case in point. As Katherine Blunt (2022) described in *California Burning*, failures to upgrade (or even maintain) century-old electrical infrastructure crashed head on into an expanding population who built in windswept areas parched by steadily worsening drought. Fires, often ignited by electrical equipment, were practically guaranteed to destroy many billions of dollars of residential and commercial property. And they have, across the state. Blunt reported the results of investigations implicating Pacific Gas & Electric's involvement in the northern California Camp fire (the company eventually pled guilty to 84 counts of involuntary manslaughter in the case [Romo, 2020]). Southern California Edison has acknowledged that its electrical equipment may have been associated with the start of the Eaton and Hurst fires in Los Angeles (Seal, 2025). Adequately fixing the problem is a job too big for a shareholder-owned electric utility, and probably also too big for the government of one of our largest and richest states. So, it never gets done. Stories like this will be told and retold all over the world in the years to come.

The financially better off among us may be shielded from a great deal of the ensuing hardship, at least for a few generations. At the same time, millions or even billions of less fortunate people may eventually show up on our doorstep demanding to join us or otherwise be compensated for the loss of their livelihoods. Western nations have already seen what more isolated calamities, such as an exceptionally intense drought (followed by civil war) like that which plagued Syria from 2006 to 2024, can do to stimulate migration. Far right political parties all over Europe, from France and the Netherlands to Italy, Germany, Austria and Hungary, have won newfound support from a growing number of anti-immigrant voters. One German worker recently made this explicit in an interview with *The Wall Street Journal*: "I haven't read the AfD's economic program and I don't care. I'll still vote for them, because of immigration." (Benoit, 2025).

If entire nations find themselves unable to grow crops, they *will* move somewhere else in numbers too great to repel with force. Climate migrants may even turn out to be our friends and neighbors. Communities all over the United States, from California and Louisiana to Florida and North Carolina, have been reduced to hollow shells, or even forced to relocate en masse, in the wake of fire and flood damage so chronic and devastating that private, and even public, insurance markets can no longer afford to cover the losses (Bittle, 2023). Economic migration in general is so politically explosive in the United States that, at the time of this writing, plans are being made to deport tens of thousands of unwanted people to the site of the 9/11 era Guantánamo Bay prison (Parti et al., 2025) in Cuba. A derelict facility originally meant to hold 200 will now be expanded to secure 30,000 people —without lawyers, external oversight, or accountability (Sullivan, 2025; Yale Law School, 2021). Guantánamo Bay is the very definition of a concentration camp, a new chapter in America's history of displacing the inconvenient in our midst that began with the Indian Removal Act and has continued with the internment of Japanese Americans during World War II and Operation Wetback during the Eisenhower Administration. We will need far more compassionate ways to address future waves of climate migration. A report by Zurich Insurance Group estimates that 1.2 billion climate migrants could emerge by 2050 (McAllister, 2024). If you are not otherwise concerned with climate change, perhaps this looming human catastrophe is enough of a reason to support action to reduce its impact.

It is important to emphasize that I do not claim that any given low carbon energy technology or policy is, *a priori*, impossible to implement. We can—for the time being—build just about anything if we are willing to put enough money and material resources into the project. Indeed, that is precisely what we have done—in an especially alarming fashion—since 1945: mineral extraction of all kinds has exploded to satisfy the demands of modern consumer society. Many of our "miracle" innovations are nothing more than the brute force application of geologically concentrated energy. A prime example is the success of twentieth century agriculture, due primarily to the widespread adoption of huge quantities of ammonia-derived fertilizers synthesized from cheap natural gas. This triumph has (temporarily?) put to rest the Malthusian/Ricardian argument that population growth would be inexorably limited by the availability of natural resources. We have instead come to believe that technology can and will solve every problem. Having replaced appeals to the favor of the gods with stubborn scientific progress, societies are now unwilling to accept any limits on their potential for material growth. Many reasonable and thoughtful people have made a compelling case for the opinion that our resource endowments are so large in aggregate that there is no chance that we will deplete any of them in our efforts to abate global climate change. I agree in principle: we will not consume the very last molecule of fossil hydrocarbon fuel, the very last atom of lithium in the Earth's crust, or —with all due respect to Dr. Seuss—the very last Truffula Tree. What we *are* well on our way to doing, however, is making the extraction of a growing number of essential resources so difficult and expensive that we will soon be forced to choose among uses for them—particularly when those uses are intended to satisfy the needs of billions of additional people. It should be self-evident that we cannot, as individuals, as nations, or even as a species, do everything we want or need to do, at the exact time of our choosing, at a cost that leaves us free to pursue whatever other aims we might desire in the future. The foregoing would suggest a wild sense of entitlement and flies in the face of physics and economics. This seemingly innate urge for instant gratification is the genesis of economist Milton Friedman's admonition that there is no such thing as a free lunch. The universe may be vast, but the Earth is a tiny, fixed mass of material located somewhere in the middle of that universe. Not only do we face real limits on the amount of anything we wish to consume, but the processes we engineer to transform raw materials into usable products inevitably alter the surrounding environment. The choices we make when designing sustainable, low carbon energy systems will have unintended consequences that will constrain our other social and economic options. With that in mind, it is crucial for the public and for policymakers to have a clear understanding of the tradeoffs and uncertainties that lie dormant in the actions we take to build new energy production and distribution technologies. What we choose *to* do must necessarily also entail choosing what we will *not* do—and not only in terms of energy consumption, but also in terms of water, other competing categories of infrastructure spending, national defense, and social welfare.

American mythology is rich and imaginative. We are deeply enchanted by our (short!) history. A sense of boundless opportunity and optimism pervades our popular culture. Mariana Mazzucato (2018), the political economist, likes to point out that the stories we tell ourselves affect our perceptions of value and the decisions we make about what is important. Classical American stories are sweeping and expansive—we are constantly on the move, chasing the western sunset, the architects of global progress and prosperity, with bountiful resources and the ingenuity to make nature do our bidding. But the stories we tell ourselves about technology can negatively affect the decisions we make about how to plan for our future. It is not that we should be pessimistic or cynical, but rather that we should be realistic. Even the stories we tell about the darkest periods of our past are often imbued with the idea that, with enough work and boldness, we can overcome any obstacle. This attitude has led us to believe that we can invent or hustle our way through anything, and that we are destined to achieve and acquire—each generation more than the last. Thus, when confronted with the challenge of climate change, our instinct is to build a better mousetrap, to outsmart the planet, to take whatever resources we need from any place we can find them, and to pursue ever higher levels of material consumption. But we cannot have everything we want, whenever we want it, at zero cost. Deep down most of us know this, but to acknowledge it feels like an admission of weakness. It feels downright un-American.

Today's increasingly sophisticated global consumers know what they have and do not have, as well as what they must do to acquire more. Poor people will not tolerate being told that they can never have the things we (their wealthier brothers and sisters) do because the way we got to where we are today was, in hindsight, terribly misguided. How the people of the developing world go about improving their lifestyles matters to all of us. As of now, they are still banking heavily on the power of coal, a substance that they are well aware is causing increasingly alarming levels of pollution. They choose to continue along this path because it is, all other things equal, better than abject poverty. Coming to terms with the ramifications of climate change will also require the more fortunate among us to make hard choices. We have neither the time, money, nor expertise to conduct an energy revolution that is sustainable, secure, and cheap all at the same time. This means we must set our expectations accordingly. For the first time in history, we are going to be forced to develop *less* convenient solutions to energy (and water and material waste) problems. We are slowly coming to grips with the fact that Earth's technically and economically usable resource endowments are finite. Meanwhile, we are also struggling with the need to accept responsibility for the impacts of past consumption on future generations. A legitimately just and sustainable clean energy transition means paying for pollution where we create it. The air, the water, and our other common resources are not, and never were, free to abuse. Nevertheless, nothing we do can prevent us from passing on a large amount of humanity's environmental debt to our descendants and imploring them to embrace the same mission strategy with their heirs. Adapting to a world of finite resources will become a permanent, global obligation, with any tangible benefit accruing only decades or centuries from now.

This message will meet with strong opposition from large corporations and other powerful interests who wish to convey the sentiment that we can have our cake and eat it too. Such sentiments are never helpful. Far more valuable would be a realistic assessment of where and how we can sacrifice to enable a transition to clean sources of energy while minimizing the unavoidable pain of doing so. We need to replace more dense, more convenient, and more efficient energy sources with more diffuse, more difficult to store and transport, and less efficient ones. This is true not only because the former are being used up much faster than Nature can replenish them, but also because uninhibited exploitation of the former has done massive harm to the environment. You can begin to gain an appreciation for the cost of this replacement by looking at the gaps between design specifications for contemporary products and the conditions for their use expected in the future. Engineers design to the demands of consumers: cost constraints typically prevent designers from choosing materials or construction methods more robust than those that meet a customer's minimum needs. A recent extreme heat event in the UK provides a perfect case study. In July of 2022 British temperatures rose as high as 38C (100.4F), a level that was previously almost unheard of. Flights from London Luton Airport and RAF Brize Norton had to be halted because portions of the runways had melted (Guardian, 2022). British airfield designers had no reasonable expectation decades ago that operating temperatures would ever exceed long-established historical patterns. Think of this as being conceptually similar to the Y2K problem, only much more widely dispersed and lacking a single solution. Redesigning the world's software to store dates using four digits for the year rather than two was expensive, but once we recognized the problem, fixing it quickly became routine. We will see more varied and unexpected design gaps as average temperatures, flood risks, and other climatic variables shift. Some nations, like the UK, will be better able to cope than others, but the cost will be high. You can't just reprogram an airport runway to melt at higher temperatures; it must be rebuilt.

Some climate induced changes will be even more disruptive. Several major European rivers will soon be impassable by barges for at least some part of each year. Transport of goods by river is still an important economic activity. If shipment over waterways becomes unreliable, then new logistical systems will have to be developed: more roads, bridges, rail, fuel storage depots, and the like. Assuming that the lowest cost shipping options are currently being used, making these needed changes will raise prices for consumers, perhaps by a great amount. Europe is already expected to need an additional $32 trillion in clean energy investment to meet its 2050 decarbonization goals. The world will have to spend a further $21.5 trillion on electrification alone (Bloomberg NEF, 2023b) (on top of everything else that urgently clamors for our attention) to fully meet demands for clean, reliable, secure, and affordable power. For perspective, the size of the world's economy as measured by GDP was $96.53 trillion in 2021 (World Bank, 2023b). Meanwhile, global military spending in 2021 was roughly $2 trillion (CIA, 2023). Canada acknowledged that it will "never" be able to meet NATO spending targets (Panetta, 2023); its current military budget is set to reach $51 billion by 2027, short of the $70 billion NATO has requested (Brewster, 2023). Half of European NATO members similarly struggle to budget for their contribution of 2 percent of GDP. The borders of Europe are being challenged as we speak. Ukraine is not currently a NATO member, but it might as well be in terms of the financial commitment being shouldered by the alliance for its defense. Clean energy spending will only add to governments' financial obligations. National security, healthcare, education, and other concerns will not pause or dissipate while we address climate change.

At the time of this writing, United States national debt stands at around $36.2 trillion (US Treasury, 2025) and Congress is once again in the throes of one of its perennial standoffs (Neuman et al., 2023) regarding lifting the limit on borrowing more. Raising the debt ceiling yet again will be one of the Trump Administration's first and most urgent second-term tasks. Former Federal Energy Regulatory Commission Commissioner Nora Mead Brownell summarized the nation's energy fiscal challenges in a May 2023 interview (Lewis, 2023):

> Well, we need to build more power lines. According to the US Department of Energy, we need to build about 47,000 miles of lines to meet the demand of the energy transition. But we're dreaming if we think that's going to happen in time to meet climate goals. Transmission lines are expensive, face considerable local opposition, and are next to impossible to permit.

Brownell went on to explain that plans to work around the constraints on new construction were primarily related to adjusting the models used to rate the capacity of power lines. The idea is that more accurate real-time transmission data can allow grid operators to safely exceed existing limits on power line capacity. In other words, we can't afford to build more capacity, and we will instead just push our existing infrastructure to the edge of its physical limits. Surely it is evident that, if we cannot resolve routine fiscal disagreements on the order of tens of billions of dollars, the probability of marshaling trillions of dollars to invest in the electric grid is nearly zero. We are tacitly counting on unrealistic levels of future spending to upgrade infrastructure while we patch the current system and hope for the best. Whether one views this problem through the lens of climate adaptation, wildfire prevention, or data center requirements, greater cooperation—along with a sober realization that American consumption cannot rise without bound—must prevail.

This book is an attempt to help prepare society for the tradeoffs and compromises that we will eventually and inevitably have to accept in the absence of infinite time, unlimited budgets, or miraculous technologies. The first step toward solving any problem is to acknowledge that it exists. The job of convincing people to do so seems only to grow more difficult as time goes on. Commenting on this rightward shift in global politics, Ursula Münch, Director of the German Academy for Political Education, wryly observed, "It turns out people [in the West] value their own jobs more than whether some islands are going to sink into the ocean" (Benoit et al., 2024). By understanding how and why humans have used energy to build civilizations, we can start to appreciate which advances have been most meaningful and essential. Only then can we design climate solutions for everyone's future that maximize our collective quality of life within the practical and indisputable constraints that our world imposes upon us.

CHAPTER 1
HOW MUCH ENERGY DO WE REALLY NEED?

I could have written this book on napkins saved from restaurants with a fountain pen. I didn't. Instead, I used a laptop computer equipped with word processing software. There's a very good reason for that: it's a whole lot easier to draft, edit, research, and reconstruct text digitally than it is to revise paper documents using a pen (or a typewriter, for that matter). I know this because I'm old enough to have done all those things. Research from home, office, or a local coffee shop using the internet is also far more convenient than having to cloister oneself in libraries for weeks or months on end. Again, I know because I spent my twenties in graduate school when in-person library research was still a daily reality.

Many things in this world are easier with modern technology, and technology is an insatiable consumer of energy. Total annual worldwide energy use is equivalent to the manual labor of more than 734 billion virtual people. That equivalent labor serves a world of only eight billion actual people (Heinberg and Fridley, 2016). This is why wars are fought over (and fortunes are made from) energy. It is impossible for the average American alive today to imagine what day-to-day existence would be like in the absence of instantly available, unlimited, ready-to-consume energy. The closest most of us ever get to experiencing energy poverty is the inconvenience of a brief power outage.

Blackouts are annoying, but they usually represent nothing more than a temporary interruption of our activities.

It is often said that water is life, and that is true, but energy is civilization. Access to energy is the difference between the nasty, brutish, and short life described by Thomas Hobbes and a life filled with art, industry, and leisure. Conflict among nations, whether economic, diplomatic, or military, increasingly occurs to establish control over energy resources. Japan's attack on Pearl Harbor in 1941 was a direct response to a U.S. embargo imposed on the energy-isolated nation, a measure that itself was in direct response to Japanese seizure of oil-rich provinces in Southeast Asia (Yergin, 1991a). The U.S. government conservatively spent $81 billion in 2018 to protect global oil supplies even during a period of relative domestic peace (leaving aside the cost of wars in Iraq and Afghanistan) (DiChristopher, 2023).

A great deal of armed conflict has been waged chiefly to protect civil society's ability to consume cheap goods. Couched in those terms, it becomes a moral imperative to carefully weigh the value of material consumption against its human costs. The natural world fares much worse. We literally destroy everything we come into contact with: forests; lakes; rivers; grasslands; animal species we never even get around to naming until after they are extinct. Even entire mountains are ground flat for the coal or metal inside them. It has been suggested, only partially in jest, that humans name places after the nature we eliminate to build upon them (just drive through any suburban housing development for examples). The atmosphere and oceans were long deemed so vast that our waste would disappear and be harmlessly diluted. We now understand—even when we don't always want to admit it—that that is not at all the case. Bleached coral reefs, perennial ocean dead zones the size of small countries, and a warming planet now finally bear witness to the scale of contemporary pollution, pollution that we can definitively trace to energy usage.

THE ROAD TO A HIGH ENERGY SOCIETY

Primitive human life had certain advantages. Perhaps the greatest was freedom from the nagging constraints of time. Karl Marx noted that the humble clock was among the first and most necessary requirements of the Industrial Age, remarking: "What, without the clock, would be a period in which the value of the commodity, and therefore the labour time necessary for its production, are the decisive factor?" (Marx and Engels, 2010). Without some way to accurately estimate the rate at which work was done, measurements of labor and productivity could not be made. Machines could not be relentlessly redesigned to minimize the former and maximize the latter; hence, the elevated status of the clock. Machines were socially transformative because they replaced the labor of dozens, hundreds, or even thousands of people. Tasks that might have taken years could suddenly be done in days or even hours —if you could supply enough energy to run the machines. Time is money, and energy buys time.

Thomas Newcomen's first steam engine is among the most famous of the early industrial machines. Its existence depended crucially on the availability of cheap coal. This pioneering design used the heat from burning coal to boil water into steam. The steam raised a piston that was connected to a rocker beam. On the other end of the rocker beam was a pump shaft. The downward pump stroke was achieved by injecting cold water into the cylinder containing the piston, thereby condensing the steam and causing the piston to fall (National Museums Scotland, 2023). Many improvements were eventually made to enhance the efficiency of steam-powered machinery, but the need for additional energy to operate machines of all kinds only increased as their numbers grew, and their usefulness became indispensable to every industry. Thus began humanity's addiction to fossil fuels.

Before our ability to access abundant, secure, and affordable energy resources, human time may have been less rigorously scheduled, but our lives were also less rich in many other ways. In the prehistoric past, before people were able to fully exploit even the simplest and most readily available energy contained in wood, wind and water, we were limited to the power available from our muscles and those of other animals. Engineers often speak in terms of energy density, the amount of energy that can be usefully extracted from a single unit of a resource. Energy dense resources are convenient because they require less time, money, and space to store and use. Anyone who remembers the difference in usability between a cellular phone designed in the 1980s and the latest iPhone can appreciate the concept of resource density: cellular phone batteries are smaller, lighter, and last much longer than they used to, because today's batteries store much more energy in the same size package than older ones could.

Here's another way to think about energy density. Imagine a world in which all energy is contained in glass marbles. There are only two kinds of marbles in our imaginary world, high energy red marbles, and zero energy blue marbles. The problem is that there are one thousand blue marbles for every red marble, so it is challenging to collect enough red marbles to power society in this world. The marbles rain down from outer space at a rate of only about a dozen each year. If you stand outside with a bucket, it will take you a very long time to capture a usable supply of red marbles, and you will have to dispose of all the blue marbles you don't need. However, there is another way to collect red marbles. It turns out that red marbles are heavier than blue marbles, so they sink into rivers and lakes while the blue marbles float on top. If you can build a machine to scoop red marbles from the bottom of a river or lake, you will gather a lot of energy very quickly. But once the river bottom or lakebed has been harvested, you're out of luck until the gradual rain of interstellar marbles slowly replenishes it, a process that takes many, many years.

This imaginary world of energy marbles is not all that different conceptually from what happens to create deposits of fossil fuels on our own planet. Take coal for example. The sun shines down on plants, which grow through the process of photosynthesis, storing the energy in their leaves and stalks. Once these plants die, they are gradually buried over millions of years under dust and sand where they decompose and are chemically transformed into hydrocarbons. Dead plants gradually decay into peat, which decays into lignite, which—given enough time and pressure—will eventually become coal. Coal is a much more energy dense fuel than crop waste or wood, the latter of which contain large amounts of nonflammable water. By comparison, the solar energy that allows for the growth of those plants is almost immeasurably diffuse (the opposite of dense). Trying to capture solar energy directly is like trying to harvest rare red marbles in a world filled with useless blue ones: it is expensive, time consuming, and inefficient. It should, therefore, come as no surprise that the rate of human technological development increased exponentially once we figured out how to use coal. This is wonderful, so long as the coal lasts, but it won't for very much longer, all things considered. Our Western calendar, calibrated to events that took place roughly two thousand years ago, will likely not survive to celebrate a Y3K if our societies are forced to depend on affordable fossil fuel for another thousand years. This fuel can be stretched—at best—a few more centuries assuming anything like current rates of use. And, sadly, energy scarcity is a constraint independent of the problems associated with human-induced climate change. Scarcity is not limited to energy resources, either. We often fail to realize that our habits of unbridled metal extraction work to frustrate attempts to replace coal, oil, and gas with "renewable" alternatives. Aluminum, gallium, and the many other inorganic elements of the periodic table (which are indispensable components of wind turbines, batteries, and solar panels) never did grow on trees.

Vaclav Smil (2017) offers a detailed accounting of prehistoric energy use in *Energy and Civilization: A History*. Early human populations were limited in size primarily by the net energy gain attainable from hunting, gathering, or agriculture. The largest and most successful hunter-gatherer groups established permanent settlements in Alaska and the Pacific Northwest thanks to the enormous energy returns they earned from hunting whales and salmon, respectively. Smil also makes the fascinating observation that Pacific Northwest salmon fishing communities turned to hunting other marine and land animals only after their growing populations needed more raw materials to make additional clothing, bedding, and hunting equipment. Net energy gain from food production was otherwise slow to advance, with very little progress over millennia. Only under extreme stress of overpopulation, drought, or persistent crop failure would civilizations resort to cultivating additional lands or labor-intensive farming enhancements. Animal power was gradually substituted for human effort. But this did not appreciably increase the number of people who could be fed from an acre of land, because more powerful animals consumed increasing fractions of the annual harvest as feed. Nevertheless, freedom from the most arduous tasks allowed each farmer to manage greater amounts of land, leading to increased cultivation and the ability to support occupational specialization. Specialization led to innovation, which in time led to the mass production of crude oil and development of the internal combustion engine. These two crucial ingredients allowed humanity to finally escape from the limitations of animal-powered agriculture.

The peak of innovation in preindustrial agriculture was reached by approximately 1900, just before gasoline powered trucks and tractors revolutionized food production. At that time, riding steel "gang" plows connected in rows of up to ten or more, pulled by dozens of horses, could harvest a hectare of wheat in mere minutes. That feat represented, compared to the hours required at the beginning of the nineteenth century, a nearly twentyfold increase in the net energy available to farmers. This advance in energy gain is illustrated in Figure 1-1 beginning with the Roman Empire in 200 and culminating with the Imperial Valley of California in 1900. Increasing the number of animals available to work a given acre of land simultaneously was an early way of increasing energy density.

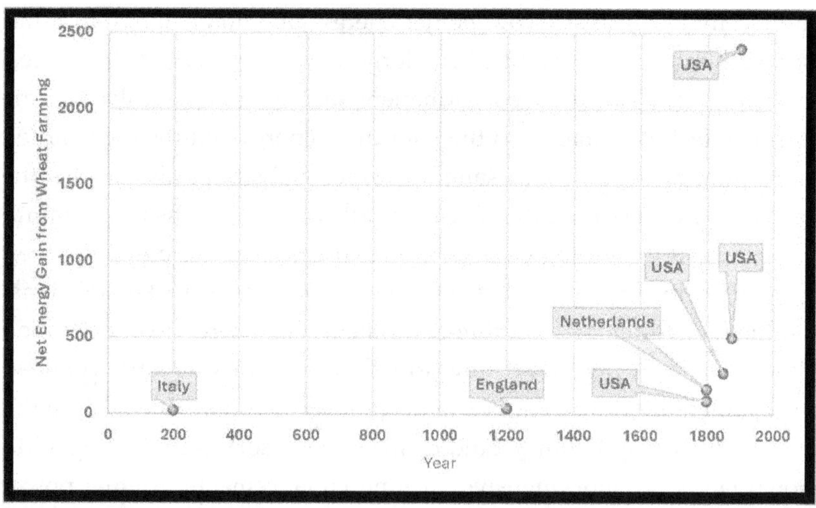

Figure 1-1. (Source: Smil)

Layton (2008) developed a method for quantifying the density of all energy resources that can account for every technology from solar radiation and wind power to nuclear fission and the combustion of fossil fuels. Layton's method can assign an energy-per-square-foot value to any land area that includes the aggregate availability of all resources in a particular location. For every cubic meter of resource volume, the amount of energy available from wind power on a moderately windy day is about one million times that available from solar power, and an average human can deliver about one thousand times as much power per unit volume as wind energy. In comparison, fossil fuels are billions of times more energy dense than any of these other resources. The energy density of oil and gas truly transformed humanity's ability to apply machine power to the construction of buildings, roads and dams, and manufactured products, not to mention the treatment of municipal and industrial wastes. Oil and gas are the reddest of red marbles.

If solar and wind have such low energy densities, why are we spending so much money on them and betting so heavily on their ability to contribute to a low carbon future? Wind and solar may not be dense energy sources, but they are practically inexhaustible. Fans of the movie *Office Space* will remember that rounding down financial transactions to the next lowest penny can result in a huge payout if repeated millions of times. The raw amount of solar radiation reaching Earth from the sun is immense and unceasing. The catch is that it is diffuse, and we struggle to capture enough of the incoming penny-sized pieces of energy to make good use of it. We are certainly getting better at doing so, but the cost of materials and land is, and will always be, high by comparison. Wind is an indirect form of solar energy: uneven heating of the Earth's surface creates volumes of warmer air in various locations at various times. Warming air causes it to rise, leaving lower pressure areas behind. Air from higher pressure locations rushes in to fill the voids, and it is those pressure driven flows that we experience as wind. Wind speeds are higher and more consistent at higher altitudes and over the oceans where there are fewer obstacles (trees, buildings, etc.) to disrupt airflow. Larger turbines can capture greater amounts of wind energy, but at growing costs of materials and land.

Resource	Energy Density (J/m³)
Solar (US Average)	0.0000015
Geothermal	0.05
Low Speed Tidal	0.5
Wind	7
High Speed Tidal	50
Human	1,000
Natural Gas	40,000,000
Gasoline	10,000,000,000
Coal	22,500,000,000
Crude Oil	45,000,000,000

Table 1-1. (Source: Layton, 2008)

The incredibly dense fossil energy sources we have come to know and love (or hate) are concentrated in fixed quantities, at least in terms of timescales that are of any concern to human civilizations. As of 2019 the world consumed roughly one hundred million barrels of oil per day (Arezki and Nysveen, 2023). Nature created every drop of this millions of years ago—before dinosaurs walked the Earth—producing it through the infinitesimally slow processes of geologic pressurization and chemical transformation (US EIA, 2023b). In this way, we see the tradeoff: energy density is an attribute won the hard way over vast timescales and is not something we can take for granted or assume we can readily replace. If we are going to solve the twin problems of resource depletion and atmospheric carbon pollution, we will have to learn to be satisfied with harvesting diffuse, low density energy sources in the future. This will be neither convenient nor cheap. It is, therefore, easy to see why fossil fuels have become so indispensable to modern life when viewed through the lens of energy density as illustrated in Table 1-1.

ENERGY TRANSITIONS IN BRIEF

The shift from wood to coal that sparked the Industrial Revolution was humanity's first significant energy transition, but it did not occur overnight as illustrated by Figure 1-2.

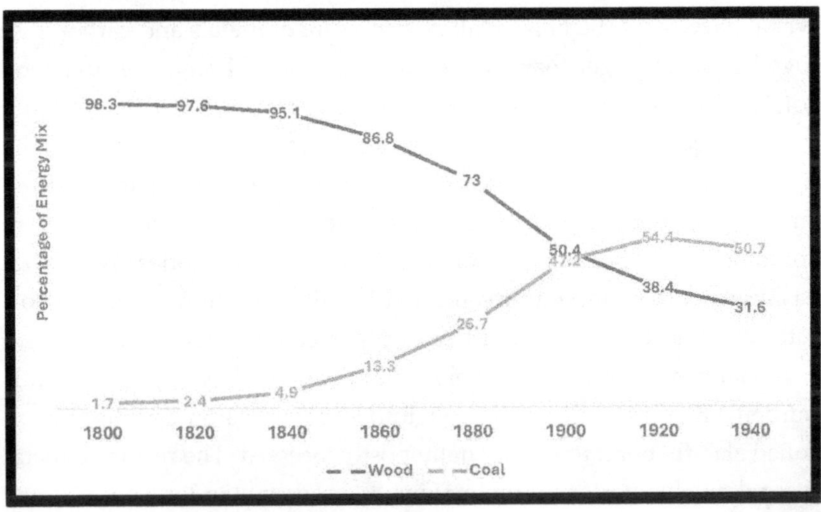

Figure 1-2. (Source: World Economic Forum, 2022)

Note that the fraction of coal use begins to taper off during the interwar period between 1920 and 1940 as the dominance of oil accelerates. Note further that we still use plenty of coal and wood as energy sources, despite having "transitioned" to the use of higher density fuels. An energy transition never actually ends; rather, successive transitions blend into each other. (There are also still a nontrivial number of farm animals working in the world's fields and on the streets of cities.) We never completely replace any source of energy. We just add more sources to satisfy the ever-expanding requirements of technological progress.

The apex of preindustrial agriculture relied not only on horsepower, but also on the availability of inexpensive steel plows. Steelmaking was a nexus where the limits of traditional farming met the beginnings of industrial energy consumption. Tilling fields is an exercise in overcoming friction to displace wet, heavy soils and deeply rooted grasses or weeds. The difference between a plowshare made of wood, one made of rough cast iron, and one made of steel is immense in terms of productivity, but also in terms of energy consumption. (Steelmaking is still one of our most energy intensive and productivity enabling industries.) Access to increasingly dense energy sources facilitated the manufacture of the steel farm implements needed to optimize wheat harvesting. Steel is an alloy, a mixture of metals and carbon that together are stronger than iron alone. Too little carbon makes the alloy soft, and too much carbon makes the alloy brittle. High temperature heat is required to melt iron ore and to remove impurities from the mixture as it is being processed. The Iron Age first began with the smelting of iron ore in clay furnaces, where wood and charcoal were common fuels. Primitive smelters, known as bloomeries, could not easily reach the temperatures needed to fully melt iron without labor-intensive injection of oxygen from leather bellows (waterwheels were used later to mechanize operation of the bellows, reducing this burden). Blooms were crude conglomerations of iron and impurities called slag that had to be manually post-processed. The resulting metal was relatively inferior for toolmaking by modern standards.

Coal eventually supplanted charcoal, particularly after coking was invented. Heating pulverized coal in a low oxygen environment removes volatile gases, tar, and other impurities, leaving behind a nearly pure solid carbon fuel called coke. Steel was obtained from smelted ore by removing excess carbon from the coke and iron mixture. Another advantage of coke over charcoal is that it contains fewer impurities which, if absorbed by the iron, would change its physical properties in unwanted ways. The switch to coke from charcoal also slowed the increasingly rapid pace of deforestation that was afflicting England and eased competition for home heating fuel and lumber. Coke is also more structurally rigid than charcoal. Using coke as a fuel enabled larger furnaces to be built, which increased iron yields. One of the first truly successful iron decarbonizing technologies was the Bessemer process. This innovation involved blowing a blast of hot air through the molten metal to oxidize impurities into slag, which sank to the bottom of the furnace. Crude carbon steel was then skimmed off the top and further refined into finished products. High quality steel can be as much as ten times stronger than pure iron. It is also harder and can be made more resistant to corrosion by blending in additives such as chromium. Steel tools reduced the human energy output required in agriculture by a substantial margin.

With more free time, families began turning their attention to other pursuits. A problem remained in that, without electricity, darkness posed an impenetrable barrier. We are all familiar with the occasional challenge of reading by candlelight, but the day has long passed when citizens of industrialized nations were limited by the setting sun. Oil lamps, used at least since the ninth century, were an initial improvement over candles (Sheridan, 2023). Whale oil was prized as lamp fuel because it burned brightly with a clear flame and little smoke. But by the mid-nineteenth century overhunting threatened to seriously constrain supplies. Kerosene was a potential illumination alternative and could be derived from coal. However, the advent of the oil age truly commoditized kerosene lamps. Yergin (1991b) relates in his epic history of the oil industry *The Prize* how kerosene distilled from crude oil was hailed as "the light of the age." Developing economies all over the world have since replicated this familiar pattern of replacing biomass with marketed fuels, traversing what has been termed an "energy ladder" (van der Kroon et al., 2011). The choice to supplant one fuel source with another is often a matter of increasing national income, although it can conflict with a nation's desire for energy self-sufficiency. Despite this potentially problematic reliance on foreign markets, energy choice offers people the chance to use the latest commercial appliances that consume a specific fuel. Kerosene has given billions of people a source of clean-burning, portable, safe light that extended their ability to study, work, and enjoy leisure activities after sunset. Even before the automobile, the oil industry first transformed society by giving people that most precious of commodities—more time.

During the earliest stages of oil exploration, gasoline was little more than a byproduct. It was of no particular use as lamp oil, for cooking, or for home heating. However, with the need for remote communities to import more and more of their daily necessities, the slow pace of animal-assisted transportation soon revealed itself to be an increasingly troublesome bottleneck. The price of progress is often paid in reduced self-sufficiency, and people accepted the bargain: more oil, less labor, more commerce. Crude oil itself, as the name implies, requires significant amounts of post-production processing to unlock its energy potential. Early oil refineries relied exclusively on the most basic of these processes, fractional distillation. The core concept of distillation is relatively easy to understand. Crude oil is not a single chemical substance, but rather a mixture of hundreds, or even thousands, of different compounds, mostly hydrocarbons—chains of carbon and hydrogen atoms arranged in various forms. This mixture is boiled in what amounts to a giant kettle, whereupon the compounds vaporize in succession from lightest to heaviest.

Another example of distillation in action is the making of whiskey. Sour corn mash is essentially a mixture of water and alcohol. Alcohol boils at a lower temperature than water, so the vapor that rises to the top of a still is richer in alcohol than in water. By capturing this separated vapor and cooling it, liquid whiskey can be drained into a jar. This same fundamental process is used to separate gasoline, kerosene, and many other useful products from crude oil, albeit using more elaborate equipment. Unfortunately, the amount of gasoline naturally obtainable from a barrel of crude oil using distillation alone is rather small. This led Standard Oil Company of Indiana (Heitmann, 2023) to invent what is known as "thermal cracking," a process of breaking large hydrocarbon molecules, such as those that might otherwise be used as asphalt or heavy fuel oil, into smaller hydrocarbon molecules like the octane we need to make gasoline. As gasoline demand grew, even more innovative methods (catalytic cracking, hydrocracking) of breaking large, unwieldy hydrocarbons into smaller, more useful pieces were developed to increase the fuel yield from crude oil. Abundant and affordable gasoline connected distant communities and accelerated the trend of buying, instead of making, many of the things that people—especially rural people—used.

ELECTRICITY AND THE BIRTH OF CONSUMERISM

The invention of vapor compression air conditioning technology did not, in and of itself, lead to the electrification of society, but one could argue that air conditioning was electricity's first killer app. Without it, the rapid development of the American South after the end of World War II would not have occurred (Badger and Blinder, 2023). Air conditioning made summer suburban commutes to sweltering offices bearable, giving a massive boost to economic productivity and quality of life. More than just lowering temperatures, modern air conditioners control humidity, an aspect of the technology that was historically even more important for industrial applications than cooling (Harford, 2023). Annual air conditioner sales doubled in China between 2008 and 2013 to sixty million units, and the fraction of homes worldwide with air conditioners is expected to rise from 13 percent in 2015 to 70 percent by 2100 (Tweed, 2015). Since the most cooling is needed on the hottest summer days, air conditioners are designed and installed to meet the highest expected demand. As a consequence, the electric grid is designed in exactly the same manner: large enough to provide maximum cooling during peak summer heat waves. Such a system is necessarily overbuilt for much of the rest of its operational life, and thus is terribly inefficient. It is also expensive. French economist Jean Pisani-Ferry has estimated that the average French family would spend 44 percent of its annual disposable income to purchase a state-of-the-art electric heat pump (and a further 120 percent of its annual disposable income for an electric car) (Ip, 2023). In the United States, a heat pump large enough to condition the air in a twelve hundred square foot home can easily cost $8,000, or much more if optional features such as a variable speed electric motor are selected.

Climate control was just the tip of the proverbial iceberg. Electrification played a starring role in giving birth to consumer society in all its aspects. Widespread electrification in the United States during the first half of the twentieth century was aided by the ambitious hydroelectric damming projects financed by the Roosevelt Administration as part of the New Deal. People welcomed cheap electricity, which let them dispense with many of their monotonous housekeeping tasks and replace them with electric appliances. But the transition was not automatic: the Rural Electrification Administration had to send instructors out into the countryside to teach people how to use their newfound freedom (Wallace, Jr., 2023). It didn't take long for people to realize that plugging in a tabletop mixer, clothes washing machine, or dishwasher was superior to hours of manual labor. Gasoline may have brought communities closer together, but electricity altered the way we worked, played, and lived, to an extent that makes it by far the most transformative human invention of all time. Nothing else since, including any of our vaunted computational marvels like satellite communications, the internet, television, or industrial automation, can exist without uninterruptible supplies of electricity. Electricity is as necessary to technology as water is to organic life.

Leisure time meant increased consumption, consumption spurred innovation, and innovation required additional and more reliable sources of energy. The result was a positive feedback loop: Cheap consumer products and the time to use them sparked more demand, which was fed by correspondingly cheap and convenient energy. These new sources of energy were sought with increasing urgency in all corners of the globe. In the minds of the public—and especially in the minds of policymakers—these rapid changes translated into higher living standards. As economic growth and geopolitical power were (and are) measured in no small part by output (gross domestic product) and control over global trade, the feedback loop expanded, implicitly and explicitly, to include national security imperatives. Examples of the interdependence of rapid economic growth and increased energy exploration abound, from the automotive sector to agriculture, aviation, entertainment, and everything between. One of the easiest illustrations to visualize is the emergence of natural gas as an economically vital energy source and its role in the production of plastics, first for military purposes and then later as an indispensable element of daily living.

Inexpensive plastic products had been enabling increased consumerism since the turn of the twentieth century (Freinkel, 2011). Demand for the raw materials from which plastics are made was a key enabler of what had been a large, but largely wasted, source of energy:

natural gas. Without a firm long-term market, natural gas cannot economically be utilized because it requires dedicated transport and storage networks. Exploding uses for electricity coincided with new markets for home heating fuel and civilian redirection of wartime plastics manufacturing capacity. All of these factors intersected with America's booming oil industry to increase the supply of and demand for gas. More miles of natural gas pipeline were built between 1945 and 1965 than at any time before or since (Joyce, 2014). The nationwide availability of natural gas, coupled with the Depression-era growth in hydroelectric power production, completed a powerful set of energy resource options for Americans that had begun first with coal and continued with oil. This energy bonanza increased the nation's living standards beyond anything that the world had ever known—or may ever know again.

Natural gas proved equally well suited to electricity generation, and electricity demand soon soared again with the advent of the computing age. Energy consumption by early computer networks was large, yet frugal by today's standards. When I was in college "the cloud" was a computer known by the moniker VAX (Virtual Address eXtension) that was housed in the basement of a building in the center of campus. Its (her?) name was Helen. A series of "dumb terminals" (called such because they had no internally installed software of their own) were tethered to the VAX by long cables. All files we created were saved on Helen's central hard drive and periodically backed up on reels of tape. To create, view, or print files, we would sit in one of the rooms full of terminals. There were no mobile devices or websites, and the incentive to spend excess time beyond that required to complete class assignments was limited. Fast forward to 2025 when software companies are accused of deliberately making their products addictive for children and where billions of people carry powerful computers in their pockets, have a second one in a backpack or briefcase, and perhaps a third, fourth, or fifth at home. It isn't really the energy required to use these small computers *per se* that is of concern, but rather the energy required to manage all the data they generate. The energy needed to keep the electronics in huge data centers cool enough to operate properly is staggering. Cooling data centers required 200 terawatt-hours of energy in 2022 (IEA, 2022b). The number of internet users worldwide doubled between 2010 and 2022, and demand from data transmission and data centers together comprise 2 percent to 3 percent of global energy consumption (Kamiya, 2022). Data center energy consumption is expected to increase to as much as 12 percent of total U.S. electricity by 2028 (US DOE, 2024). This spike in demand is a direct consequence of the sudden rise of artificial intelligence (AI), cryptocurrency, and other computationally intensive, "big data" industries. The trend toward broader economic "datafication" will almost certainly be replicated worldwide.

Humanity's earliest machines were rudimentary tools designed to confer a mechanical advantage to muscle power: levers, pulleys, wedges. Once the energy required to construct them had been expended, these tools lasted for years, perhaps even generations, without need for upgrading or even significant maintenance in many cases. Today's machines are often obsolete as soon as they reach store shelves, and they consume growing amounts of energy throughout the course of their useful lives. Those lives grow ever shorter as consumers strive to purchase replacements having the latest aesthetic designs and features. Energy and the associated extraction of raw materials for manufacturing have become so inexpensive that millions of people can afford to buy more goods in a year than their grandparents acquired in a lifetime. Every gadget we buy promises to do more for us and to enrich our lives in ways we had, more often than not, never bothered to consider. I still watch movies, listen to music, and read books. Is it easier to find content today than in 1980? Sure. Is the entertainment experience itself really any different or better? Sometimes yes, other times no. Is it all worth the extra energy consumption? That's an important question to answer in terms of climate change and resource depletion, and it is a question that deserves even greater thought when looking at some of the most energy intensive future technologies, like AI, cryptocurrency, and space tourism.

PROGRESS OR PROFLIGACY?

Even the most ordinary and reliable mechanisms in our lives, from doorbells and locks to dishwashers and refrigerators, are expected to become "smart." We are told that this is for our convenience as well as a necessary part of the energy transition, yet it leaves every purchase suddenly equipped to consume *more* energy and become more quickly outdated. In some cases, digital control of appliances is expected to increase the stability of energy demand throughout the day by, for instance, starting a load of your laundry while you're asleep and not using your cinema size television, vacuum cleaner, or gaming console. Appliance automation nevertheless represents an additional energy load, both in terms of the power they consume and the materials and labor used to build these increasingly complex devices. How much energy do we really need to live a civilized existence? Is there a happy medium somewhere between the drudgery of preindustrial society and an ever-increasing acceleration of activities we might choose to defer, forgo, or do the old-fashioned way without sacrificing all that much real convenience? Are the locks I open by inserting a key in a door or my refrigerator that lacks wi-fi holding me back in some way I just can't understand? I don't believe so, but perhaps I haven't yet been targeted by the right advertisements.

What does a stable energy economy look like? The notion of an energy transition implies that there is an end point—that the process will be completed at some point in time and that we will reach a point of equilibrium. At that point we would use a core mix of energy technologies that are different from those that came before. But the future of energy production and consumption is more likely to come to resemble a dynamic state of continual adaptation. As populations grow and demand higher standards of living, and as the pace of resource extraction increases, we will be forced to imagine ever more sophisticated, flexible, and costly ways to optimize and transform raw materials and energy delivery systems. Population growth is decelerating. By 2100 it may return nearly to preindustrial rates, with further increases concentrated largely in Africa (Piketty, 2014). Even at that relatively plateaued level of population, however, material consumption will remain orders of magnitude greater than it was before 1700.

Moreover, clean energy technology will, at least for many years to come, be constructed using existing sources of less clean energy. With that in mind, prioritization of the ways in which we will use that energy budget seems prudent. In a world where we had all the copper we could ever want to build new transmission wires, all of the lithium and nickel needed to build batteries forever, and all of the money and labor and political will to accomplish everything else we could possibly desire, there might be less need to debate the relative merits among various uses of electricity. Here on our real Earth, our budget is finite, and our window of time to act is rather short. We cannot risk allocating critical resources without a plan. People who live in drought-prone areas are already told how they may use water because it is simply too precious to waste—even if a few wealthy people could afford to do so. Rationing materials is necessary for sustainable living, and that requires public policy choices to be made, such as growing lawns versus lettuce.

What is, and is not, worth spending our remaining easily accessible stores of fossil energy to create? This is a controversial topic. Attempts to legislate or regulate economic activity are rightly fraught with concerns over property rights and civil liberties. Unfortunately, the tension between wants and needs goes hand in hand with scarcity. Few are likely to dispute that energy for pumping water, heating and cooling, basic voice and video communication, and effective transportation should rise to the top of the list in a world of reduced energy availability. The value of other products is less clear. A detailed analysis of AI, its capabilities, and the ethics associated with its use is outside the scope of this book. Nevertheless, its deployment will have a measurable impact on the speed with which a clean energy transition can be accomplished, and therefore on the amount of low carbon energy that will be available to people for doing everything else they want and need to do. Most of those people will not live in the few countries that broadly use AI applications in the near future, but they *will* live in countries which host the associated computer servers, power plants, and cooling systems needed to sustain our new robot colleagues.

How much energy and carbon will AI cost us? Nobody really knows, because the industry is unregulated, much of its technology is a closely guarded trade secret, and its profitability remains an open question. An internet search using the ChatGPT product developed by OpenAI uses ten times as much electricity as a conventional Google search (UNRICWE, 2025). And the water needed to cool the electrical components in AI data centers worldwide is expected to exceed the annual water usage of Denmark by 2027. Some newer models may be more energy efficient, as evidenced by the recent performance of Deep-Seek, an OpenAI competitor that claims to have drastically reduced its "training" (calibration) cost with respect to similarly capable products. It is, at the time of this writing, still an open question as to whether this cost (and energy) savings is being achieved—in whole or in part—through reliance on existing, expensively trained models (Schechner, 2025).

In any case, it is easy to imagine the proliferation of AI models creating a significant growth in demand for energy resources. A study conducted at the VU Amsterdam School of Business and Economics estimated that the AI industry could be using as much electricity as an entire nation the size of the Netherlands by 2027 (Kleinman and Vallance, 2023). A more recent estimate, noting that data centers hosting OpenAI's GPT-4 required 30 MW of electricity (approximately as much power as is consumed by thirty Walmart stores), predicts that by 2030 the largest AI models could consume 5GW—the amount of electricity currently consumed by Manhattan (Hiller, 2025a). Some AI applications may deliver new medical insights, enable resource conservation in other industries, or otherwise justify themselves. Others may have the potential to add value yet could be suitably replaced by less energy intensive and easier to regulate alternatives. Autonomous vehicles come to mind. Self-driving features are coming to be inseparably associated with the world's expected transition to electric vehicles, and this is deeply puzzling. There is nothing about an electric motor and battery drive mechanism that requires or even really suggests that the human operator be eliminated. If urban congestion, inattentive drivers, and traffic fatalities are truly the problems we claim they are, why are more of us not taking the bus? A billion autonomous vehicles, each relying on algorithms to stay on the road and avoid hitting each other, people, or animals, could consume as much additional power as all existing data centers (Zewe, 2023). Many AI applications will be frivolous or even harmful. For an example of the latter, imagine a chatbot that counsels children while archiving their most intimate thoughts as data that could potentially be sold on the open market (Jargon, 2025).

Cryptocurrency could grow into another huge energy sink. Digital currencies not issued or backed by governments rely, like AI, on complex mathematical algorithms. This structure is used to certify that individual "coins" are unique and to place limits on the amount of currency that can be "mined" (the analog of a printing press in the physical money world) to maintain long term user confidence in their monetary value. Mining cryptocurrency involves using a computer to solve mathematical problems, the reward for which is a small unit of new cryptocurrency. Competition among miners to be first to solve these problems determines who earns more, setting up a digital arms race that deploys powerful computers bidding for ever larger fractions of available electrical generation capacity. The computing power needed to manage cryptocurrencies is immense. Estimates of the power consumed by cryptocurrency mining in 2022 were 150 TWh per year, equal to the electricity used by Argentina (Hinsdale, 2022). Add to this energy bill the burden of additional bank oversight, money laundering prevention, and other safeguards to enable the secure use of these new financial products. Are entirely new forms of money worth diverting time, materials, and electricity away from core decarbonization efforts on such a scale? In the quest for a zero-carbon economy, discussions about the regulation of emergent technologies must include assessments of their energy impacts weighed against the social benefits of any less resource-intensive alternatives.

RESOURCE COMPETITION IS NOT GOING AWAY

Geopolitical competition over mineral resources, either to perpetuate the energy status quo or to dominate the world's energy future, could imperil global progress toward climate change mitigation. The architect of German unification, Otto von Bismarck, believed that world affairs could not be managed by speeches or democratic decisions, but rather were destined to be settled in iron and blood. To date, I would say that he has, by and large, been vindicated. Military planners and politicians are acutely aware that they cannot create or coerce a greater geographic diversity of mineral deposits, wind currents, or solar radiative intensities. Instead, we fight over what exists. And no empire, including the British, French, Spanish, Japanese, or—lately—the United States of America, has managed to enforce a stable global trading system through deployment of a blue water navy. Each has tried, and the cost eventually became prohibitive. China's recent claim of sovereignty over the South China Sea is the most recent effort of this kind, and it will likewise siphon massive resources from that nation's economy. Meanwhile, great power countries have also long had the habit of making energy producing nations perennial scenes of revolution, bloodshed, and diversion of those nations' leaders' attention from improving their own governments. This is largely the modern history of the Middle East, a perverse engineering of borders and coups for the benefit of Western petroleum companies. Outbreak of war in Ukraine in 2022 that severely disrupted energy flows to Europe was accompanied by warnings from Russia, one of the world's largest producers of petroleum products, that nuclear weapons would be a justifiable means of protecting its economic and territorial integrity. China currently controls a great majority of the production and processing of scarce elements needed to produce electric vehicles, wind turbines, solar panels, and other clean energy products. The Chinese government has used its mineral dominance as a geopolitical cudgel, forcing multinational corporations to transfer more advanced technology to China in exchange for access to raw materials (Pitron, 2020). Deposits of many key minerals, such as cobalt, nickel, and lithium, are heavily concentrated in just one or a few nations, such as the Democratic Republic of Congo, Indonesia, and Bolivia, that are often poor and/or politically unstable.

A race among great powers to dominate oil-rich nations drove the economic colonization of weaker regions, from the Arabian Peninsula to West Africa and Southeast Asia. This trend now threatens to spread to the Arctic as the polar icecap melts and year-round sea lanes open in the far north. Russia has recently practiced firing cruise missiles near Norway, Finland, and Sweden in response to earlier Arctic wargames conducted by NATO countries, and the region is seeing intensified commercial competition for the immense untapped resource deposits lying under the fast-melting ice. (Kiss et al., 2025). The world cannot afford to repeat its reckless gamesmanship over energy and mineral reserves in an era of hypersonic missiles and weapons of mass destruction. There can be no energy stability in such a geopolitical environment. Nor can the energy transition be predicated on hoarding a stockpile of oil replacement minerals at gunpoint. The losers of that competition will inevitably be forced to continue emitting greenhouse gases, a scenario that will not benefit the winners, with whom they share the same atmosphere and climate. We should instead learn to better adapt our consumption to the limits of what we can reasonably produce without resorting to violence. Conservation, efficiency, and flexibility are elements long in aspiration, but short in supply, among the world's energy policies and markets.

A stable future energy economy will have to involve more economic actors of all sizes and incorporate new ways of shepherding investment among a mix of technologies that are best poised to decarbonize the world rapidly and robustly. This mix may or may not include the buzziest ideas or the ones promoted by the most politically connected executives. For its part, the role of government must be to remove barriers to interconnection of emerging systems and to develop shared infrastructure that can grow and transform with minimal disruption as the technology portfolio adapts to changes in resource constraints, geopolitical stresses, and lifestyles. Climate change may turn out to be an opportunity to reimagine some of the huge, unwieldy structures we have been using to deliver essential commodities. The World Bank has estimated that, by 2050, one hundred forty-three million people will migrate in response to resource scarcity caused by climate change (Podesta, 2019). It is unclear where all these people will go, but they *will* go somewhere. And climate-induced migration is not only a transnational phenomenon. The Stanford Woods Institute for the Environment reports that evidence already exists for interstate migration within the United States by people who are increasingly concerned with the growing danger of unpredictable wildfires, floods, and other natural disasters (Jordan, 2021). Migration represents another form of resource competition—competition over land.

Land competition is a constraint on development, but it can also positively influence the design of future energy systems. Electrical transmission infrastructure was designed for an era in which very large generating plants were built in locations remote from the customers they served (von Meier, 2006). Power plants have historically been hazardous, polluting, and in need of immense supplies of fuel that must be delivered at all times of the day and night. We did not want them, as the saying goes, in our backyards. Today we can make a credible case for more distributed generation of electricity, hydrogen, and other forms of energy storage, rather than blindly replicating continent-spanning networks of pipelines and wires. Allowing small communities and neighborhoods to be more energy self-sufficient helps to ameliorate some of the uncertainty and cost associated with the moving target of a shifting population. Self-sufficient energy modules can still scale to provide services to large populations through automated interconnectivity. This strategy, known as Autonomous Energy Systems (NREL, 2023a) not only helps to reduce the risk of cascading power failures spreading to large geographic regions (an increasing problem with today's electric grid), but can also help to smooth out the fluctuations inherent in renewable energy systems comprised of diverse sources. Linking distributed energy islands into networks that can replicate the performance of centralized systems is a challenge, but one that is being met with algorithms developed at the National Renewable Energy Laboratory. The small town of Basalt, Colorado, has been equipped with technology to share energy from solar panels, batteries, heaters, and electric vehicles installed by individual homeowners across the community of 4,170 (O'Neil, 2022). The approach, called Network Optimized Distributed Energy Systems (NODES) was part of a collaboration with the U.S. Department of Energy's Advanced Projects Research Agency. This venture is an excellent example of the inclusive design of supply and distribution networks.

Yet an energy transition that redistributes public infrastructure onto private property could also do a great deal of harm. It is virtually beyond living memory in this nation that anyone has ever had to give a thought to whether moving to any place, from the smallest town to the largest city, would put at risk their ability to acquire safe and reliable access to electricity (and water). Many American neighborhoods are already blessed or cursed by their proximity to particular school districts. (That problem, too, is being addressed by many policymakers through privatization [charter schools].) Without universal access to high quality electricity, poor neighborhoods will only grow less livable. A home is also the primary asset used to hold retirement savings for a great many workers. A chain reaction of residential neighborhood decay, beginning with lower public utility investment, and continuing with rising uninsurability, and falling property values, would create a swift and severe drag on economic productivity, the financial stability of seniors, and overall civic cohesiveness. We need to think broadly about such things before blithely designing quick fix technologies that could generate widespread social ills.

The alternative, unfortunately, is to completely refurbish and replace infrastructure components on a scale that the world has never seen, a scale that makes the work of building Egypt's pyramids appear trivial. One description of the effort needed to enable a futuristic, green electricity generation and transmission system is "a power grid long enough to reach the sun" (Bloomberg NEF, 2023a). This same report goes on to note that 1000 GW of solar electricity generation capacity and 500 GW of wind capacity is stranded, waiting to be connected to the grids of the United States and of Europe. Only 21 percent of new clean energy projects proposed by developers ever actually transmit power over the U.S. grid (Osaka, 2023d). Power lines cross hundreds of miles to reach end users, a distance over which large numbers of landowners predictably protest their intrusion. Streamlining the permitting process for new construction worries almost everyone other than project developers, even environmentalists, who fear collateral damage to ecosystems or that new power will be generated from conventional fuels rather than clean alternatives.

WE CAN'T DO IT ALL

Since 1950, infrastructure spending in the United States has typically fluctuated between 0.75 percent and 1.25 percent of GDP (Tomer et al., 2021). Large increases notably occurred during the mid-1960s and late 1970s as delayed responses to passage of the Interstate Highway Act in 1956 and the Clean Water Act in 1972 (Davis, 2017). The Infrastructure Investment and Jobs Act passed in 2021 is projected to result in $1.2 trillion in future spending, with $550 billion of that amount an increase over baseline budgets. However, more than $300 billion of that $550 billion amount is reserved for non-electric transportation and broadband projects, and another $63.3 billion will be used to upgrade water infrastructure. Can we realistically expect, in the face of the myriad other fiscal challenges governments confront—including national security, health and pension support for aging populations, transportation, water, and other critical infrastructure—that historically unprecedented clean energy spending goals will be met?

We will have to innovate, yes, but also to compromise. Assumptions about infrastructure progress that prevailed in the nineteenth and early twentieth centuries never were, and never will be, sustainable. We are no longer advancing against a wild limitless frontier, but rather attempting to replace entrenched systems in the face of rising debt. Shifting populations may make infrastructure capacity calculations obsolete long before many of these multi-billion-dollar projects reach the end of their useful lives, which can be thirty to forty years or even longer. Consider that between 1950 and 1990, while New York City retained its place as the largest urban area in the United States, Houston jumped from fourteenth to fourth, San Diego from thirty-first to sixth and Phoenix from ninety-ninth to ninth (U.S. Census Bureau, 1998). Even to the extent that such trends are predictable, what are planners to do? When cities lose their luster (Baltimore fell from sixth to twelfth, Pittsburgh from twelfth to fortieth, and Buffalo fifteenth to fiftieth over the same period), a lack of public investment can cause urban decline to become a self-fulfilling prophesy. On the other hand, overbuilding can leave half-finished concrete shells blighting the landscape and spell ruin for investors. A famous example of this is the fate of five nuclear power plants in the State of Washington. Of the five, only one was completed. Two of the unfinished plants resulted in one of the largest bond defaults in history. The remaining two incurred billions of dollars in financing costs, forcing the Bonneville Power Administration to raise electricity rates 418 percent from 1978 to 1984 (NPCC, 2024). Inaccurate electricity load forecasts were at least partially to blame.

Smaller, modular energy systems that can be more easily resized to match dynamically changing customer pools are one option. Modularity creates its own problems, though. Rooftop solar power production is a good example of the unintended consequences of new technologies. The more that distributed generation displaces power from large, centralized plants, the less revenue major utilities earn to maintain the grid that is still needed to stitch all these tiny power producers and consumers together (Bakke, 2016). Market innovation will thus be as important as the physical upgrades needed to transform our energy delivery systems. Costs for shared infrastructure will have to be equitably allocated, and proper incentives for conservation— rather than for relentless growth—must become part of the cumbersome and slow-to-adapt electricity rate structures managed by regulators.

How much of everything do we *really* need? I haven't owned a four-wheeled vehicle in nearly ten years. My wife and I share a car. I also have a motorcycle that I can use to run small errands or take trips that don't involve large luggage loads, and I can get 60 or more miles per gallon on the highway. Would I contribute more to sustainability by purchasing an electric vehicle? At this stage of my life, it is unclear. Should I expect to take a transcontinental flight once a year, or is that a luxury that is incompatible with emissions reduction targets if enjoyed by billions of people just like me? Is watching television at home on a device large enough to credibly function as a movie theater screen truly necessary, even if I can afford to do it? Energy technology is constantly advancing to meet the growing needs of societies striving to improve their living standards. Yet, as of 2021, there were still 800 million people without any access to electricity (IEA, 2021a), a number that has dropped by approximately 50 percent over the past two decades. The desires of people in lower income countries to consume cannot be separated from the global race to decarbonize. Technological constraints will never allow billions of humans to enjoy a *sustainable* lifestyle at the level of early twenty-first century middle-class American civilization indefinitely. The rate of resource depletion can be made faster or slower, accompanied by more or less waste. But we are ultimately constrained by nature to live within the bounds of the First (mandatory conservation of energy) and Second (inevitable production of entropy) Laws of Thermodynamics. More will be said of this in future chapters, but the point is that everything we choose to build consumes a finite resource in the absolute sense. We transform matter and energy at rates far more rapid than natural replenishment processes can ever cope with, and the gap is destined to widen over time.

Over the course of the next few chapters I encourage the reader to reflect on both the potential and the limitations inherent in each of the core pillars of the energy transition: electrification, carbon capture, the hydrogen economy, and recycling. Who will benefit and who stands to lose from reorganizations of the commodity trading markets that must occur to facilitate accurate, global pricing of carbon? How much, and which kinds of materials, can realistically be recycled? How many of the key raw materials needed to produce clean energy can be extracted without irrevocably damaging other aspects of the environment? The answers to these questions will determine how rapidly, equitably, and democratically societies can "go green," and perhaps will determine whether they can do so at all.

CHAPTER 2

WHY DOES THIS TRANSITION HAVE TO BE SO HARD?

There were few deeply entrenched business interests standing in the way of delivering modern energy to the vast American West during the 1930s and 1940s. People like my grandmother who had washed clothes and dishes laboriously by hand quickly became accustomed to using machines as electric power lines steadily made their way into small towns and isolated farms as part of the New Deal era's Rural Electrification Act. (Nevertheless, grandma kept her washboard propped up in the corner of the laundry room decades after the electric washing machine arrived. Just in case.) The pace of infrastructure improvement was rapid and, in comparison to the present day, relatively uncontroversial. There were certainly land use conflicts and safety concerns, and not everyone was initially convinced of the value of new technology. But there were no large, complex energy infrastructures or industries needing to be kept in constant operation during our western frontier's transition from kerosene and horsepower to electricity. To borrow a bit of contemporary terminology, the arrival of electricity into an area previously dependent on primitive forms of energy was not disruptive. The more pervasive an infrastructure network becomes, the more consumer habits and the wider economy must bend to align with it. Economic alignment along infrastructures is an investment that, over time, becomes increasingly difficult to alter. The stickiness of technology makes infrastructure replacement even more difficult than its construction cost alone might dictate.

The pre-electric era was labor, not capital, intensive. Replacing labor with technology offered immediate rewards and few drawbacks. To replace electricity today would instantly render useless billions of appliances, tools, games, and communication devices in which people have invested cumulative lifetimes of earnings. Even if whatever hypothetical energy source brought into use as an electricity substitute was cheaper and more sustainable, replacement of all the things that can only run on electricity would be an insurmountable barrier for most consumers. No one is suggesting that we replace electricity; just the opposite. Yet, aspects of the clean energy transition do involve the wholesale replacement of huge technology networks, or at least their repurposing or gradual obsolescence and abandonment. The need or desire to retain familiar patterns of fuel or product usage will likely have as much to say about the design of a sustainable future as emissions calculations or energy security, at least for many decades to come.

Early energy transitions had less disruption to contend with than the one we are presently engaged in. The nineteenth century mechanization of skilled labor did create social tension, famously exemplified by the Luddite protests in England. Workers smashed machines and set fire to factories where owners refused to enforce labor standards, especially those regulating wage rates and credentialing (apprenticeship) (Conniff, 2011). The Luddites' fight was not really against the machines *per se*, nor the replacement of wood with coal. More importantly, textile mill owners were not attempting to displace a nationwide infrastructure of commodity extraction, production, transportation, trading, and consumption technologies. They could essentially isolate the impacts of mechanical innovations within their own shops: steam powered machinery could be installed without seriously jeopardizing existing delivery contracts or the need to replace costly existing equipment. Contrast this with the difficulty, expense, and potential for chaos of having to first decommission a coal-burning power plant and then replace it with a wind farm and backup system needed to smooth out fluctuations in power generation caused by wind speed variations.

Clean transportation poses even larger challenges. Continent-spanning networks of consumer-facing refueling systems must be replaced —essentially all at once—if new vehicle energy sources are to be made viable for intercity, let alone interstate, trips. These changes will affect the livelihoods of millions of people who operate the existing infrastructure: service station owners, tanker truck drivers, manufacturers of vehicles, generators and other appliances that use existing fuels, and of course the producers and traders of existing energy commodities. For this reason, the time required to implement today's clean energy transition will far exceed the transition from wood to coal, or even from coal to oil. Each succeeding energy replacement has created larger and more extensive webs of infrastructure to support it. These webs are collections of long-lived assets which drive the economies of many nations—particularly the nations most responsible for carbon emissions. Smil (2017) has concluded that the time a society needs to integrate a new primary energy source has been relatively constant, between fifty and seventy-five years. He based his analysis on transitions between traditional biomass (wood) and coal, between coal and crude oil, and between crude oil and natural gas. Yet, the share of replacement in each case has been smaller, being over 50 percent for coal, less than 50 percent for crude oil, and around 25 percent for natural gas. It is unlikely that these replacement shares will be matched by any combination of nuclear (fission or fusion), hydropower, hydrogen, geothermal, or biofuels anytime soon due to economic and land use constraints. However, the combined electricity share of wind and solar could reach or even exceed 50 percent by the end of the twenty-first century, perhaps sooner. Hence the early realization among many planners and policymakers that electrification is the first and best overall path to global decarbonization.

To emphasize the general challenge of infrastructure switching, consider a few of the most significant of our existing energy network investments.

FLUID STORAGE AND TRANSPORT

The effective usage of liquid or gaseous fuels at regional scale requires pipelines for transport and extensive surface or subsurface facilities for storage. Storage of oil or natural gas in subsurface salt caverns is a mature technology, but one that depends on the presence of relatively scarce geological features. Tank farms occupy large land footprints. Pipelines of all types face increasing opposition from landowners, environmentalists, and the public at large. Nevertheless, development of networks of oil and natural gas pipelines to serve continental scale customer bases is now commonplace. However, in almost all areas of the world, these pipelines must cross multiple political jurisdictions. Geopolitical risk of energy supply disruption is as much of a challenge now as is ensuring technical integrity of the infrastructure itself. Nowhere has this been more conclusively demonstrated than in the supply of German natural gas by Russia. Germany has been an early mover in efforts to clean and stabilize its energy economy. Former Chancellor Angela Merkel took concrete steps to accelerate the replacement of coal and nuclear power in Germany with Russian gas, a decision that was—at the time—expected to reduce the German nation's carbon intensity of GDP and enhance energy supply reliability (AP, 2022). Technology adoption at scale demands a great deal of faith and ongoing commitment.

The physical and technical constraints that apply to natural gas transportation and storage will also apply to hydrogen, along with the added complications of chemical reactivity and increased explosivity. Hydrogen will also cause ordinary steel to become brittle. Replacing ordinary steel with more suitable alloys makes hydrogen more expensive to transport and store than natural gas. Liquefaction of hydrogen is possible only at extremely low temperatures (-253°C or -423.4°F), and) and is thus more expensive than similar processes used to liquefy natural gas (-162°C or -259.6°F). Hydrogen has long been produced in industrial quantities. A network of fit-for-purpose hydrogen pipelines and storage facilities crisscrosses the industrial region spanning the Houston Ship Channel and the Louisiana Chemical Corridor. It has operated reliably for decades. This network was built at great expense because of hydrogen's central role as an input to the oil refining and ammonia production processes, both of which are clustered in the same area. Other transportation methods that sidestep hydrogen's physical complications involve transforming pure hydrogen into easier to store liquid carrier chemicals, such as ammonia or methanol. The carrier chemical is either disassembled back into hydrogen at the receiving point, or else is used as a fuel in its own right: both are feasible, at a cost.

Some environmental advocates warn that the pace of a clean energy transition may be slower if so-called "bridge fuels" such as natural gas are deployed in the near-to-medium term. Fuel bridging should not necessarily be construed as a failure. Although conventional natural gas infrastructure cannot always be used without modification to transport hydrogen, it is amenable to retrofitting with suitable materials. Avoiding coal in favor of gas already reduces direct emissions (those from combustion, not accounting for fugitive methane losses) by about 50 percent. Gas turbine power plants can be designed to run on a variety of cleaner fuels, including hydrogen, ammonia, and methanol. Infrastructure plans that are affordable, realistic, and chart a clear path toward energy decarbonization, security, and reliability are also what the world's poorer regions need immediately. Hydrogen-compatible natural gas pipelines, compressor stations, and power plants can be a flexible long-term clean energy policy option when coupled with economic incentives to move gradually from gray (derived from natural gas) to blue (gray plus carbon capture) to green (derived from carbon free electricity) hydrogen. Dual fuel (hydrogen/natural gas) technology reduces the risk and costs of eliminating proven networks, markets, and machines; it also leverages storability benefits of gaseous energy at scales that are unachievable with battery backup. We are accustomed to storing (summer) and withdrawing (winter) enough natural gas to manage the seasonal energy needs of the entire continental United States. The largest battery farms cannot provide more than a few hours of local electricity backup.

CENTRALIZED ELECTRICITY PRODUCTION: AN ESSENTIAL PUBLIC GOOD

The most intuitive and earliest way of building electrical networks was to simply connect producers to consumers with a wire, allowing current to flow steadily at a constant rate. This approach, aptly named direct current, was effective but inefficient. Just as friction along a pipeline rapidly causes water pressure to fall, requiring pumps at regular intervals to boost fluid energy, the electrical energy in direct current dissipates quickly as it encounters the resistance in transmission wires. Direct current could not originally be transmitted more than a few miles economically, and the water pumping analogy is not exact, because to boost the energy in a direct current line meant generating higher current at the source and building larger wires. Modern high voltage direct current (HVDC) lines have overcome many of these early limitations (Larson, 2018). It is possible that, with the better technology we have today, if we were to start over and build a brand new electric grid from scratch, direct current might be more competitive for power transmission. Thomas Edison, who created the first successful commercial electric grid in the United States, relied on direct current. Edison's business plan involved replicating small, local power generation plants all over New York City at intervals of approximately one per square mile (Bakke, 2016). The proposition became increasingly unsuitable for a growing urban customer base, and it proved to be totally unacceptable for delivering electricity between cities.

Alternating current allowed for longer distance transmission with acceptable loss rates. It soon became common to build large power plants in remote areas close to coal mines or along rail lines. Smoke from rural coal combustion dissipated instead of choking cities, and the massive fuel deliveries needed to produce electricity at a regional scale could be centralized. The overall shape, or topography, of the network did not really change. Electricity was designed to flow in a single direction, from a small number of central generating facilities outward in an expanding circle to large numbers of customers, most of whom used only a tiny amount of power. All the equipment needed to deliver power was designed in accordance with this fundamental principle of centralized, outward flow. Large, high-capacity wires were strung between plants and distribution substations, smaller, medium-capacity wires led to each neighborhood, and the smallest, low-capacity wires served individual homes. Plant engineers balanced power by monitoring voltages and currents at key locations in real time and making adjustments at a local or subregional level. No one considered that the day might come when every home could have its own rooftop solar power plant and an electric car in the garage serving as a battery of last resort for grid balancing. It isn't just that power now needs to flow "backwards," but that the number of simultaneous producers and consumers of power is expected to multiply beyond the capabilities of engineers to effectively manage on their own. One potential solution relies on advanced predictive software tools (NREL, 2023b). It aims to measure grid performance within microseconds and anticipate problems across statewide service areas, allowing operators to avoid many otherwise catastrophic failures.

Unfortunately, adding sophistication to the computer systems that control essential networks opens them to greater risk of disruption by cyberattack and to the brittleness (susceptibility to sudden malfunction) of real-time algorithm calculations with little or no room for error. Control systems for the nation's seventy-three hundred power plants, one hundred sixty thousand miles of high voltage lines, and millions of lower voltage distribution lines are increasingly being connected to the internet using commodity, off-the-shelf networking equipment and software. These common devices are economically attractive for system operators, but they often contain the same vulnerabilities that afflict other similarly equipped business networks (Brooks, 2023). All equipment connected to the internet then becomes part of the grid, which, in addition to core networking equipment that may be hardened against attack, will also include billions of diverse "edge" devices, from personal computers and thermostats to vehicles and refrigerators. Hackers don't necessarily have to control a power plant to wreak havoc, but instead can target the ability of people to live comfortably or hamper productivity in innumerable small ways. And since the voltage and current along every branch of this network must be balanced every second of every hour of every day, timing of automated control systems—particularly those that rely on satellite signals such as GPS—can be disrupted by a single well aimed pinpoint attack. Batteries and other system buffers can be strategically placed to dampen some effects of power fluctuations, but the buffers themselves then also become part of the grid to be supplied and controlled.

The cost of upgrading global electric infrastructure to meet all the aspirations of universal electrification advocates for transportation, climate control, industrial production, and communications is more than the world has been prepared to spend., That means compromises will inevitably have to be made. The difference between our relatively ad hoc grid that grew organically out of the simple collection of wires originating from Edison's earliest efforts and a "smart grid of the future" is stark, aptly illustrated by the following definition from a report by the International Energy Agency (2022c):

> A smart grid is an electricity network that uses digital and other advanced technologies to monitor and manage the transport of electricity from all generation sources to meet the varying electricity demands of end users. Smart grids coordinate the needs and capabilities of all generators, grid operators, end users and electricity market stakeholders to operate all parts of the system as efficiently as possible, minimising costs and environmental impacts while maximising system reliability, resilience, flexibility and stability.

In the absence of trillions of dollars in public funding, microgrids, community solar farms, and individual homesteads producing their own photovoltaic electricity will increasingly be used to connect wealthier pockets of society and to bypass those who cannot afford to join. This approach will bring us closer to the self-sustaining dwellings that were common before the development of automobiles and rural electrification. In other words, a likely side effect of decentralization is the exclusion of less affluent communities from the highest levels of utility service. We have long taken it for granted that virtually any home or apartment building in the United States has ready access to the same quality of electrical, telephone, and water service. Internet service has been a notable exception to this utility paradigm and much has been said about the consequences of the digital divide on opportunities for children who grow up on the wrong side of the fiber optic cable. The Covid 19 pandemic has been a natural experiment on the relationship between educational performance and access to technology. Schools and homes where technological skill, broadband access, and computer hardware availability were highest had a much easier time coping with the necessity of remote learning. It is not difficult to imagine that compromised access to other forms of basic utility service will also exacerbate income and wealth inequality in every society where it occurs. Spending on electric grid modernization would have to double through 2030, and be sustained thereafter, if the world intends on meeting the 2050 decarbonization goals espoused under the Paris Agreement (IEA, 2022c). Appetites for increasing government debt are falling rapidly in an environment of rising interest rates and ballooning deficits. Taxation is never popular. That leaves electric ratepayers to shoulder the burden. More of these ratepayers may be the poor left behind with aging systems as the rich build renewably powered islands for themselves.

ROADS VERSUS RAILS

Rail travel, from high-speed intercity service to urban commuter systems, is convenient and comparatively easy to electrify. It ameliorates many of the worst side effects of automobile ownership: insurance premiums; distracted and drunk drivers; regular maintenance; parking. The United States made development of the railroads a national security imperative during the 1860s. The slaughter of Native Americans, "men, women and children," by the army was considered both a necessary and acceptable measure to protect railroad construction (King, 2012b). Once completed, the transcontinental railroad facilitated the near extinction of the American bison. None of this destruction really registered as a sacrifice or a loss to the nation. And then, after all that blood and treasure, we suddenly decided we didn't need to travel by rail anymore. We had cars and the Eisenhower Interstate Highway System. Congress created Amtrak in 1971 to keep passenger rail service on life support, where it has remained in intensive care ever since. We also had airplanes. We had plenty of oil. Everything in America was bright and shiny and new and without limitations or regrets when viewed in the light of postwar glory. Even the twin energy crises of the 1970s were not enough to challenge the dominance of travel via personal automobile.

To change our collective travel mode preferences now would require something far more challenging for us (tragically) than genocide or species extermination. We would have to give up our highways and cars. Although younger generations, at least in the United States, appear to be less insistent on personal automobiles, a wholesale disruption of that travel paradigm is considered so outlandishly impossible that instead we seek to create robot cars to drive us individually along the highways. For the longest zero carbon trips, we would also have to surrender the immediacy of air travel. Reallocating limited transportation budgets from road to rail would accelerate an already visible degradation of roads and bridges that threatens public safety, increases vehicle wear and tear, lowers business productivity, and raises frustration levels. Yet, any credible plan to reduce greenhouse gas pollution in the United States must necessarily involve revitalization of our passenger rail system at the expense of airlines and automakers—even electric ones. Ironically, as big a fight as that would be, it ranks among the easiest of sustainability tasks to implement because lower carbon rail technology is proven and globally available right now.

CAPITAL FLOWS: OUR GOLDEN HANDCUFFS

Infrastructure replacement is not limited to physical plant and equipment. Our entire way of life is crucially dependent on financial and economic machinery such as stock and bond markets, banks, and pension funds. Companies' market valuations, the servicing of their debt, and the security of an aging population all demand that any changes to the ways in which we produce and consume energy and materials do not suddenly upend the payment systems people depend on. We cannot abruptly cease the operations of large portions of our existing economy as a precondition of achieving net-zero goals. The emergency Covid 19 shutdowns in 2020 did temporarily make meaningful dents in global oil consumption and carbon emissions, but at a terrible cost to the average household. And a large, across-the-board tax like the Trump Administration's reciprocal global tariffs also tends to have the same disruptive braking effect on the economy. Over a mere few days in April 2025 tariff announcements erased several trillion dollars in wealth from global markets. Much of this wealth was in the retirement accounts of seniors and the college savings accounts of parents—people who needed their money immediately. At the same time, speculative traders willing to bet on market volatility may have profited handsomely at the expense of ordinary investors (Schneid, 2025).

Yet the climate stabilization strategy known as "degrowth" proposes to do something even more disruptive than a blanket tariff policy. Degrowth advocates argue (correctly) that the continued consumption of resources in service of infinite economic growth is inherently unsustainable. Their proposed solution is to eliminate economic growth as it is customarily measured in terms of gross domestic product (GDP). Market transactions would be replaced in many cases by community ownership of resources. Consumption and competition would be replaced by a focus on well-being and personal development. If that sounds a lot like a hippie commune to you, you've hit upon the movement's biggest problem. While it is a gross and unfair oversimplification to dismiss degrowth as nothing but a wild leftist fantasy, such characterizations are a predictable byproduct of the mixing of degrowth's resource conservation and decarbonization goals with calls for culture war style social reform (Kallis et al., 2020). But, make no mistake, for degrowth to be at all feasible it would have to be implemented through wealth redistribution: The greatest economic sacrifices would be made by those most able to absorb them, rather than being spread uniformly across what is an increasingly inegalitarian society. A carefully designed program of degrowth that leads to targeted reductions in many of our most frivolous energy expenditures is almost certainly necessary for long-term sustainability.

The status quo is *not* a long-term solution. Capitalist economies today function like huge pyramid schemes. We produce goods and services to grow profits, which are then invested in more goods and services in an endless cycle. When you buy a share of stock, you expect its value to go up over time. In fact, you desperately need it to go up, because asset appreciation is the only thing that will allow you to stop working and enjoy life before you die. This is true irrespective of whether you have a defined benefit pension plan, an individual retirement account, or are counting on a government-run social insurance program. The only way to keep the economic escalator moving relentlessly upward is to have a bottomless reservoir of raw materials and labor that creates more goods for ever larger markets to keep stock prices rising. The end of profit means the end of income (which of course means the end of income tax revenue), the end of dividends, and the end of the leisure class. Under a degrowth regime, you would consume fewer goods and services. So would everyone else. As a consequence, less of everything would be produced throughout the economy. Unfortunately, production actually does need to fall eventually, because the cost to extract ever scarcer deposits of minerals and to grow food from increasingly depleted soils will eventually outstrip technology's ability to optimize them. If you do not believe that, then you are making the implicit argument that the laws of classical thermodynamics do not hold at the scale of the global economy. We will return to this topic during the discussion of physical limits on energy production.

Allowing for the end of conventional economic growth and collapsing the pyramid on which billions of people have built their expectations for a dignified retirement is a nonstarter. Citizens will riot, and nations will go to war, to stop this from happening. Universal basic income and services such as healthcare and education could be workarounds to enable a manageable decline in economic activity, but they cannot succeed with the stark wealth inequality that exists today in most developed countries. People who can buy their way out of what will become little more than a glorified distribution of food stamps will come to rule over a mass of powerless, resentful people. We have seen this before: it is called feudalism. In an economy where almost everyone earns a fixed, minimal amount per year, the cost of subsistence living will, unsurprisingly, settle at the exact value of that stipend. Any expense in excess of mere survival will incur a debt that can never be repaid, because real economic compensation for employment will fade away. This was, more or less, the lot of Southern sharecroppers after Reconstruction, and it is not living in the sense that anyone aspires to. Having all labor done by robots does not change this, because the resources to design, build, and maintain the machines will have to be allocated by someone. Those few people, an elite priesthood of technology, will not be content to share equally with the masses any more than the richest among us are today. After a few generations of universal basic income, the owners of the robots would be keeping the rest of an increasingly uneducated humanity as livestock.

Unfortunately, modern finance is not well positioned to solve all our problems either. Mathematical finance suffers from an inherent logical inconsistency because it relies on the tacit assumption that a business can exist and grow in perpetuity. The technical basis of this flaw is a construct from calculus called an infinite series of terms that is used to compute what financial modelers call discounted cash flows. New markets, new revenue, and new supplies of labor and materials are assumed always to appear if the price is right. The value of any human endeavor is predicated on this assumption of infinite growth, an evidently poor assumption. A better model of business value would embrace the realities of material constraints, which is to acknowledge the laws of mass and energy conservation. The practical market for any product is limited in size, and the amount of material, capital, and labor available to serve any market must be diverted from some other use. A given quantity of aluminum can be used to make beer cans or airplane wings, but not both at the same time. An engineer can work for Budweiser or Boeing, but not both at the same time. If money is to retain its value, a bank can spend only what is in its vault to finance a product. Banks circumvent this restriction by creating money through the very act of making loans—money today is just a promise from a computer. The banking system props up our illusion of infinite economic growth, and this fallacy clouds our judgment about the ability to sustainably implement any real-world program, including the clean energy transition.

LEAPFROGGING LEGACY SYSTEMS WITHOUT STUMBLING

Economic anthropologist Jason Hickel explains succinctly how the quest for economic growth drives energy consumption at a faster rate than we are developing clean energy production capacity. In 2019 the world was producing 8 TWh more renewable energy annually than it did in 2000, but overall energy consumption rose by 48 TWh (Hickel, 2019). To catch up, the world will have to begin producing the equivalent of 6 kWh of carbon-free electricity for every single kWh that is used anywhere for anything, all of this without coal, natural gas, or oil. Ironically, decarbonization of the developing world might be less controversial than tackling this problem at home—if we can take a page from the story of the telecommunications industry in Africa. In 2019 the Institute of Electrical and Electronics Engineers estimated that Africa was on track to have one billion mobile phones by 2022 (Strickland, 2019). Some areas of the continent were expected to adopt 5G networks directly, and very little landline infrastructure exists compared to the birthplace of the telephone, North America. Telephony in Africa is an example of leapfrogging: technology improvements that can bypass earlier generations, including their entrenched business interests and consumer habits. As a result, Africans have embraced mobile payment systems with much greater speed and enthusiasm than Americans, many of whom still write paper checks and visit ATMs to get cash.

What this means for the energy transition is that one of the easiest ways to slow climate change is to start with places that never fully carbonized. Eighty percent of the world's population rely on imported energy (IRENA, 2022). It is unrealistic and unfair for the wealthiest nations, which are responsible for starting the problem of carbon pollution, to expect billions of poorer people to accept a permanently low standard of living. Investing in the prosperity of our neighbors can work in our own self-interest as well. References are often made to the celebrated Marshall Plan, a $13.3 billion (approximately $142 billion in 2022 dollars) investment made to rebuild shattered European economies in the wake of World War II (National Archives, 2023). The

Plan was as much a defense against the advance of communism into western Europe as it was an altruistic act (Steil, 2019). Grants, not loans, for clean energy projects in the developing world would help Western companies to perfect competitive, advanced technologies, build goodwill among the world's citizens, and make tangible progress toward climate change mitigation goals.

China still burns half of the coal consumed by the world every year (Brant, 2023.) This is not because Chinese people don't care about the climate, but because coal is what they have and because they want to have easier lives than their parents and grandparents did. Many other parts of the world share similar energy resource portfolios and aspirations. It should be no surprise that China's Belt and Road Initiative (BRI) has resulted in the building of coal fired power plants in developing countries. However, of the fifty-two coal fired power plants initially planned for construction in conjunction with BRI, thirty-three were canceled or deferred, eleven were still in the planning stages, and only one was operational as of June 2021 (Normile, 2021). Paradoxically, 87 percent of the financing for overseas coal power generation comes instead from Western or Japanese lenders, whereas solar, wind and hydropower have surpassed coal as fractions of BRI investment since 2020. Chinese provinces have continued to use coal projects to stimulate local economic growth since the pandemic, but their shift toward renewable energy in development financing must be acknowledged and—more importantly—mirrored by the United States and other Western economic powers, including private sector entities.

Wealthy nations, despite having an abundance of capital, are struggling to deploy low carbon electricity generation systems at scales needed to make meaningful reductions in their greenhouse gas emission levels. In 2021 more than eighty-one hundred electricity infrastructure projects were stalled, waiting for permission to connect to the grid, up from fifty-six hundred in 2020 (Plumer, 2023). Wind and solar power plants are especially vulnerable to grid interconnection delays because their electricity production is intermittent—it fluctuates from minute to minute with changes in the weather. Baseload power, typically derived either from the combustion of coal or natural gas, or from nuclear or geothermal heat energy, provides a reliable, steady electrical load that is much easier for grid engineers to manage. Ninety-five percent of proposed electrical generation capacity in 2022 was wind, solar, or battery storage, and the amount (1250 GW) of solar and wind awaiting interconnection to the grid is approximately equal to the entire amount of generation capacity currently installed in the United States (LBNL, 2022). Europe is facing similar challenges, with wind and solar interconnection wait times in the UK between six and ten years (Plimmer, 2022). New plant permitting times have risen to five years in continental Europe (Ford, 2023). Avoiding additional sea level rise is one of the most obvious potential benefits of building clean energy infrastructure for poor nations—a benefit that is already badly needed in many U.S. coastal areas. Sea levels along U.S. coastlines are expected to rise as much between 2022 and 2052 as they did over the preceding one hundred years, resulting in damaging floods occurring more often and further inland (NOAA, 2022). We can choose to pay the price of that sea level rise in terms of coastal city rebuilding, migration of coastal populations, and disruption of economic activity. Or we can do the more useful work of decarbonizing wherever feasible before those costs come due. And, as those of us in wealthy countries become more and more aware of the true risks of climate inaction, we can use the knowledge we have gained to continue the buildout of low carbon power generation in our own backyards.

China could restructure the BRI and take the lead in providing clean energy infrastructure financing to the developing world. The BRI is China's Marshall Plan, designed to nurture trade, investment, and diplomatic links with the rest of the world. It is a conduit for the dispersion of Chinese technologies and, indirectly, cultural and political values. Originally envisioned as a reopening of the ancient Silk Road connecting East Asia with Europe, it has acquired global scope and reach. The onerous credit terms offered by Chinese companies have caused alarm in some development finance circles, with warnings that participation amounts to a debt trap (McBride et al., 2023). Moreover, China's prior emphasis on coal plant construction was unhelpful from a climate perspective. A broadened and redesigned clean energy investment program organized by leading industrialized nations, including China, for the benefit of the developing world could avoid such controversies. Such a strategy could reduce global greenhouse gas emission levels faster than piecemeal projects in host countries that must contend with the replacement of entrenched infrastructure and consumer habits.

Developing nations are eager to build clean energy infrastructure. The chronic barrier is financing to overcome the high initial costs of the projects (Papathanasiou, 2023). Even if the projects can pay back these costs over their economic lives, borrowing terms can create debt servicing burdens that compromise governments' ability to provide for their citizens' other needs. In the long term, China's BRI is expected to leave twelve of forty-three low- to medium-income participating countries vulnerable to heightened debt (Bandiera and Tsiropoulos, 2019). The world can do better. Affordable, responsible development was an initial promise of the Bretton Woods institutions: the World Bank, and the International Monetary Fund. Of late, the Western postwar order has not delivered on that promise. With the looming urgency of climate change, it is past time to refresh or replace these organizations with new, and more inclusive, leadership and more generous funding.

Sometimes, the decision to add carbon-emitting energy production isn't just a response to fuel or capital availability, but a reflection of how important fossil fuel production is to the core economy of that region. In other words, developing countries may prefer coal or gas infrastructure even in cases where they are offered money to adopt cleaner projects. Before we continue this discussion, let's review the concept of economic rent. The economist David Ricardo viewed the aristocratic landowners of eighteenth-century Britain as one of the greatest impediments to economic growth, because they collected what he asserted was unearned income from the rents they charged to tenant farmers. He viewed economic rent as unearned because it was derived from monopoly control of a scarce resource. The best farmland is always cultivated first and produces the highest crop yields. As a population grows, the additional land tilled is less productive. Crops sell at a single price in the market at harvest time, so the owner of the best land makes more profit than the owner of less productive land. This excess profit grows larger and larger as increasing amounts of less productive land are used for agriculture. This concentrates economic resources in an aristocracy, greatly enriching the owners of the best land, and their heirs, generation after generation. Rather than investing in innovation, these wealthy land barons spend the lion's share of a nation's output on luxurious living.

The modern Middle East is a region where economic rents dominate foreign and domestic policy. Leaders in Middle Eastern countries remain trapped between the need to finance government spending with hydrocarbon exports and the need to invest in alternative industries in which they will almost certainly will never enjoy the competitive advantages that made them wealthy in the past. Russia, too, relies heavily on oil production, with energy revenue routinely comprising as much as 10 percent or more of GDP. China has had a similar economic dependence on coal production, again explaining coal fired power plant construction as a leading export technology to less developed countries. Approximately half of the world's oil and gas is produced not by Saudi Arabia or other OPEC countries, but by middle-income countries such as Brazil and India with per-capita GDP between $1,036 and $12,535 (Saha et al., 2023). For comparison, the U.S. per capita GDP was $70,249 in 2021 (World Bank, 2023c). Losing their energy industries would be for many countries akin to the United States losing a major land war that eliminated a sizable portion of its territory: telling them that the world must stop burning hydrocarbons without simultaneously providing a credible pathway to other forms of industrialization and a middle-class existence is not only unfair, but it will also simply fall on deaf ears.

Despite the United States reprising its role as a world leader in oil and gas production courtesy of the shale boom—thanks to the success of hydraulic fracturing and horizontal drilling—energy remains a relatively small fraction of its highly diversified economy. Energy production in the United States is nevertheless still considered essential from a national security perspective, and the energy industry commands outsize political influence and receives large government subsidies. Energy security is another hurdle if the world is to implement comprehensive decarbonization programs. Countries gain a geopolitical advantage when they are energy self-sufficient, and still more advantage in keeping other nations energy dependent. To voluntarily cede a competitive energy advantage, whether through unilateral domestic decarbonization or the buildup of energy infrastructure abroad, will mean a real or perceived loss of international prestige and influence. Moreover, nations that receive support for such voluntary decarbonization pathways cannot be expected to trap their perhaps already struggling economies in a fresh web of debt: donor nations must recognize energy finance as crucial foreign aid that is in everyone's mutual interest. This can be a galling prospect for residents of wealthy nations who have their own local challenges to contend with. But if we choose not to make proactive climate investments in poorer parts of the world, we should be prepared to calculate the maximum local cost of climate change mitigation that we would accept in exchange for accelerating overseas decarbonization efforts. Climate change cannot be compartmentalized, fenced off, or threatened with ICBMs. Everything that happens to the atmosphere, or to the oceans, propagates along with air and water currents to every other part of the world—often in unpredictable ways—but also in ways that we can readily envision and budget for.

NOBODY WANTS AUSTERITY

Unfortunately, the best efforts of technologists, diplomats, and economists will no longer be enough to prevent damaging effects of global warming. Less warming is still preferable to more warming, and we will need to drastically curtail wasteful energy consumption to achieve anything close to a net zero economy (to slow the rate of warming) or a carbon negative economy (to actually begin to reduce warming). It was only rationing, and the lifestyle changes American families accepted (often grudgingly) during the Second World War that allowed factories to rapidly retool from producing consumer products to supporting the defense industry. Only with a similar level of austerity could we implement a comprehensive energy transition in the United States in anything less than a scale of decades to centuries. However, to prevent future warming, any new austerity program would have to last not for a mere few years, but effectively forever. Any politician who even suggests such a thing will be immediately voted out of office. The only achievable transition will thus involve a mixture of climate change prevention and climate change adaptation.

The original moonshot is also frequently used as a rallying cry by clean energy advocates: as in, if we can go to the moon, we can do anything. It has seductive appeal, but is a poor metaphor for the task ahead. The 1960s space program was one of the most visible examples of the United States recklessly spending its peace dividend. We ended World War II as the world's largest creditor nation and promptly set up the Bretton Woods system to allow us to preferentially borrow in our own currency—the only currency that was deemed convertible to gold. The moonshot, like the Vietnam War, the Atlas rocket program, and, earlier, the Manhattan Project, was essentially uncoupled from financial reality. We are now the world's largest *debtor* nation, with Congress regularly and predictably coming within days or even hours of default, unable either to cut spending or agree to borrow more. The climate crisis will be the most expensive national and global emergency in history. We are out of time and nearly out of credit, and we face much more painful choices than merely printing money to throw at the problem.

There is another mismatch associated with the moonshot metaphor. The space program was, more than anything else, a competition between the United States and the Soviet Union, a proxy conflict that was something between a professional chess tournament and the Cuban Missile Crisis. When the Soviets succeeded in launching Sputnik, we just couldn't stand the idea of communists getting the better of us: our competitive juices started flowing and we were united against an identifiable adversary. An equivalent animosity against some imagined external foe is not materializing to motivate us to tackle climate change with the same intensity (deep down, we know that it's our own damn fault). The consequence of our collective ambivalence is that governments around the world would be well advised to begin aggressively preparing for climate change adaptation, at least as much and probably more than they are doing belatedly to *prevent* global warming. We will inevitably overshoot the warming target of the Paris Agreement. Hundreds of millions of people will be forced to migrate out of the worst affected areas, both within and across national borders. Eventually there will be a panicked effort to eliminate emissions, and it may even finally serve as a catalyst for true global cooperation, but the road to that point will be rougher than almost anyone currently anticipates.

We can expect that today's energy transition will continue at a slower pace than any that came before. We are replacing dense, easy to transport and easy to store commodities such as coal and gasoline with diffuse, difficult to capture energy sources such as wind and solar radiation. Both wind and solar must then be transformed into electricity to be immediately consumed, transferred to batteries for short-term storage, converted to a longer-term storage medium (such as hydrogen or pumped water stored in a reservoir), or else dissipated. Hydrogen, if produced in meaningful amounts, will eventually compete for scarce geologic storage space with captured carbon dioxide. Most of the best hydroelectric resources have already been developed. We are contemplating infrastructure investments that would exceed anything humanity has ever constructed, while simultaneously ensuring the constant and uninterrupted reliability of everything that it would replace. At the same time, we are expanding the demand for energy to unprecedented levels as the billions of people living outside the world's richest countries clamor to achieve American style living standards without destroying what remains of the environment in the process. Land use constraints will force more of us to learn to live in close proximity to large industrial machinery. Meanwhile, we will likely have to dispense with some of our cherished habits: frequent trips to beaches halfway around the world, homes large enough to store unused possessions we can't even remember, backyard swimming pools that consume drinking water, and private transportation. So far, we have declined to do so.

CHAPTER 3
THE LAWS OF PHYSICS AND ENERGY LIMITS

was trained as a mechanical engineer, a field that, as taught in universities today, is almost impossibly broad and includes not only machine design, but also biomechanics, robotics, applied fluid dynamics, heat and mass transfer, electronics packaging, chemical thermodynamics, building climate control systems, nanotechnology, and thermal power plant design. I chose to specialize in thermal-fluid engineering, a discipline with a skill set applicable to energy systems analysis and development. Whatever your chosen specialty, you only really need to learn three things to become a mechanical engineer: conserve mass; conserve momentum; conserve energy. One might argue that to be successful in practice there is a fourth: conserve money, which is related to the minimization of entropy production. The practice of any branch of engineering is, at its core, the optimization of a process or product within a set of existing constraints. In designing energy systems, conservation laws lie at the heart of all constraints engineers face.

Before we go farther, it is useful to reflect on two words I have already used, but which will become even more important in the context of the following chapters: energy and entropy. You may think that you have a pretty solid understanding of energy, and it might surprise you to learn that science has no generally accepted definition of what energy actually *is*. One of the best thermodynamics textbooks I ever read was H. C. Van Ness's *Understanding Thermodynamics* (1983): it is a treasure. Van Ness explains, using the most pedestrian yet illustrative examples, how energy is an abstract concept. We detect and measure it only indirectly through changes in the properties of carefully demarcated portions of the universe we call "systems." Something is either in the system, or it is part of the surroundings. Over the years, we have discovered that, no matter what we do to a system (heat it up, squeeze it, haul it up a mountain, bombard it with radiation), there is a set of properties (temperature, pressure, volume) that always change together in ways which ensure that the amount of this abstract "thing" inside the system and passing through its boundaries remains perfectly conserved, or constant. In our day-to-day lives, we appreciate energy because we can easily observe many useful changes in objects around us caused by transferring it from one place to another or from one physical manifestation to another (heating our homes, propelling our cars, pumping water). The conservation of energy is known as the First Law of Thermodynamics. We accept it as universally true because it has never been disproven by experiment.

Entropy, on the other hand, is not a household name. Van Ness cleverly shows how it can be "discovered" using the same simple, methodical approach that leads one to understand that energy is conserved. There are many ways to alter the properties of a system while conserving its total energy. For example, you can let air out of a balloon all at once, or very slowly. You can raise a weighted platform that holds a spring compressed beneath it by removing all the weight (represented, let's say, by a bag of sand) at once, or by carefully removing grains of sand from the bag slowly one at a time. Why does that matter? It matters because the amount of useful work you can get out of the system is intimately related to the way, or the process, used to change its properties from one state to another. The efficiency of any process is always highest along one perfect, unique path that engineers call "reversible." Every real-world process is *irreversible;* less than perfect. The more irreversible your process is, the more of this second abstract "thing" called entropy you generate. The fact that entropy, again detected and measured only indirectly through changes in properties we can observe, always increases is known as the Second Law of

Thermodynamics. People have tried for over a century to find a way to violate it, and if you end up being the first I can virtually guarantee you a Nobel Prize. Most engineering textbooks loosely define entropy as "disorder," a nod to the statistical link between molecular behavior and classical thermodynamics. For purposes of reading this book and thinking about sustainability, imagine entropy to be closely aligned with waste or pollution. In such an analogy, the more of it you generate, the less efficient your system will be, and the more it will cost to operate, unless you can foist the pollution cost onto someone else (economists call this kind of cost transfer an "externality").

The manifestations and parameters of design constraints are always changing, so what is optimally designed today may appear terribly misguided in hindsight. The decision to use coal as fuel for the first steam engines was not a bad decision at the time. Not only did the switch from wood to coal help slow the rapid deforestation of Europe, but coal was, as we have discussed, a much more energy dense fuel that enabled more power to be generated more quickly in a more compact form. Yes, the burning of coal emitted carbon, but the atmosphere was able to absorb it. The carbon content of the atmosphere is always changing in a natural cycle that spans thousands of years. A small amount of anthropogenic carbon added to that cycle in the 1700s was akin to adding a couple of grains of salt to a glass of water—you can safely drink the water and can't taste the salt. Fast forward to 2025 and we have emptied a whole jar of salt into the glass. Such a mixture would be completely saturated, with excess solid salt accumulating at the bottom.

Thermodynamic limitations have long featured in discussions of sustainability. These limitations and their consequences form core concepts of a philosophy known variously as Earth limits, planetary boundaries, ecological economics or, applied more broadly, thermoeconomics. The idea is straightforward and intuitive: you can't get something for nothing. If you reject this, then your new theory must be able to replicate all of the explanations of classical thermodynamics as well as the ability of a closed system (the physical goods-based economy of an isolated planet, for example) to grow forever without replenishing its resource base. Many people desperately want to believe that some clever and charismatic entrepreneur will always arise ready to beat nature at its own game and wrest control of the Earth from the laws of physics. This wasn't always our attitude. Humans used to live in abject fear and awe of natural phenomena, ascribing the causes of observable events to gods they imagined to be watching and manipulating the world from some unseen place. English philosopher John Foster (2022) remarks that the Enlightenment triggered a widespread change in our perceptions of human intellectual prowess. Having awakened from the relative ignorance of the Middle Ages, Western societies were suddenly enthralled by their newfound ability to make sense of the world around them. The notion that, with a little effort, any problem could eventually be solved and any barrier overcome, came to be regarded as common sense. The pendulum has swung completely to the opposite side, whereupon our modern hubris is now as dangerous as was our ancient superstition.

The fundamentals of physics were, and are, inherently fixed. Our understanding of them continues to evolve, as does our ability to use them to our advantage. However, that knowledge does not make them subject to our will. They merely act, consistently and relentlessly, everywhere and anywhere, at every moment, whether we fully comprehend them, and whether we respect them. The fantastic advances in material transformation and consumption that have characterized the past two hundred years do not represent human mastery over nature, but rather human exploitation of nature. The bill for this expenditure has come due. McNeill and Engelke (2016) paint a compelling picture in *The Great Acceleration: An Environmental History of the Anthropocene Since 1945* of how rapidly and comprehensively we have been depleting the Earth's resources. Of the endowment of fossil fuel left to humanity during five hundred million years of past geochemical activity, between fifty and one hundred fifty million years' worth has been burned since 1950. Until 1965, virtually all this energy was consumed in North America and Europe for the benefit of one-fifth of the world's population. If the entire world had been living an American lifestyle throughout the past seven decades, we would have already exhausted these resources. Global freshwater usage ballooned from 580 cubic kilometers in 1900 to 1,366 cubic kilometers in 1950 and 3,900 cubic kilometers by 2011. Virtually all (98.8 percent) fresh water on Earth exists as glacial or polar ice or as groundwater (USGS, 2018), while most water for human consumption is drawn from lakes and rivers (the other 1.2 percent). World population has undergone slightly less than a fourfold increase from 1900 to 2000, whereas production of lead, iron, zinc, and copper have increased roughly seven-, twelve-, seventeen- and twenty-ninefold over the same period (Wagner and Fettweis, 2001).

At some stage, demand on our planet for every kind of resource will begin to outstrip the supply we can feasibly recover even after taking technological progress into account. The reason technology will struggle to keep up with resource extraction demand is related to entropy. We always start producing the easiest to discover and refine deposits of any resource first, because they are the least expensive to acquire. That cheapness is no accident: it has a direct link to the relatively low irreversibility of processing high quality minerals. Harken back to the heady days of the California Gold Rush, when an intrepid explorer with pickax and shovel, or perhaps a pan and set of hip waders, could sift through perhaps 300 kg of rock to recover enough pure gold to make a wedding ring (Conway, 2023). Today's gold mines are sprawling pits trawled by trucks each the size of a small building. Anywhere from 4,000 to 20,000 kg of material is excavated and discarded to recover the gold for an equivalent wedding ring nowadays. Waste equals entropy.

THE INFINITE RESERVOIR AND SUSTAINABILITY

This brings us to the concept of an infinite acting reservoir, of which Earth's atmosphere before the Industrial Revolution can serve as a good example. First picture a large plastic jug with a spout at the bottom. You may have seen these at restaurants, campgrounds, or picnics. When the container is full, and you press the button to open the valve, the rate at which liquid comes out is determined by the size of the spout. No matter how hard you press the button, the liquid will not flow any faster than the diameter of the spout will allow. An engineer would say that the valve sizing or capacity is controlling the flow.

From the perspective of the spout (and the thirsty camper), the supply of liquid is infinite—the flow is steady and strong. But what happens when the container is almost empty? The flow slows to a trickle, and you then have two choices. You can tip the container toward you, which requires you to be able to balance it carefully (and is not advisable if the liquid is red Kool-Aid). You can alternatively press harder on the button in an attempt to coax the valve open even wider, thereby increasing the diameter available for flow through the spout. This trick might work for a little while. In either case, the container is no longer acting as an infinite reservoir—the valve can "detect" the drop in liquid pressure because the limitation on flow rate is no longer the size of the spout, but rather the level of liquid remaining in the body of the container.

The Earth's atmosphere has behaved in an analogous way while being filled with carbon over the past several centuries. The atmosphere is a huge mass of fluid (mostly air), and the addition of carbon dioxide from eighteenth century coal combustion was initially unnoticeable. Early carbon dioxide emissions quietly dispersed into the larger volume of oxygen and nitrogen at a concentration that was no more noticeable than a couple of grains of salt in a large glass of water. At this initially low level, no appreciable rise in global temperature from coal burning would have occurred. The atmosphere at that time was effectively an infinitely acting reservoir, and far away from the island of Great Britain there was no way to even measure the slowly increasing concentration of carbon dioxide. As time went by, and more nations began to industrialize, the amount of carbon entering the atmosphere increased to the point that it could not be diluted enough to avoid exceeding Earth's recent historical average level. By the late twentieth century, the rise in atmospheric carbon concentration was detectable anywhere and everywhere, and the greenhouse effect of the increased carbon dioxide was changing the global climate. There is no way to permanently stabilize the amount of carbon dioxide in the atmosphere: it is always either increasing or decreasing, and we can only add to the rate at which this occurs. Absent human intervention, the rate is balanced by uptake of carbon in the oceans, in plants, and in minerals. There is no single, perfect equilibrium concentration of carbon in the atmosphere, only a range of levels within which the global climate is not rapidly being altered. The climate is a dynamic entity. It is always changing in small ways over geologic timescales to balance the emission of carbon through volcanic activity, wildfires, release from the oceans, and animal respiration with its cyclical storage in the atmosphere, oceans, and rocks, its consumption by growing plant life, and its emission by decaying organic material (Herring, 2020). When we artificially exceed these natural rates at which carbon is balanced, climate stability is impossible.

Limitations in other natural cyclical processes affect the replenishment rates of resources. Some species of pine, among the fastest growing trees, can grow as much as one or two feet per year (Arbor Day Foundation, 2023). In contrast, it takes a thousand years for a meter of bog to decompose into peat (Wildlife Trusts, 2023). Millions of years are required to further bury and compress this material under heat and pressure to coal (Geoscience Australia, 2023), or for the tiny plants and animals washed to the bottom of ancient seabeds to be compressed into crude oil or natural gas (Union of Concerned Scientists, 2023; BBC, 2023). We consume these resources at a far faster rate: global oil demand as of 2023 stands at roughly one hundred million barrels per day (US EIA, 2023b), where the natural (geologic) replacement rate is approximately 2.7 million barrels per *year* (Miller, 1996). Stated another way, we are consuming the world's existing stock of fossil fuels at a rate over 13,500 times faster than nature can regenerate it. When we talk about renewable energy, what we are really talking about is the re-establishment of a natural replenishment rate, a rate of energy harvesting that matches nature's ability to recycle it geochemically. Renewability, security, reliability, and affordability are together the aspects of sustainability, and it is sustainability, not renewability, that we must be striving for in the clean energy transition. Later we will see that, for many substances of industrial importance, especially metals, the rates at which natural processes aggregate them into concentrated, readily minable deposits are so slow that humanity cannot reasonably expect ever to recover them sustainably, any more than we can expect to wait for today's plants and microorganisms to be transformed into fresh crude oil many millions of years from now.

ENERGY CONVERSION AND EFFICIENCY

Home heating is a prime target of electrification efforts. Replacing oil-fired boilers and natural gas furnaces is, in many respects, very low hanging decarbonization fruit. Every home already uses electricity for almost everything else, and electric heat pumps are now able to function in virtually any climate. Of course, they are just as susceptible to power outages as any other electrical appliance, and a natural gas furnace paired with a portable generator remains an attractive option in the very coldest areas. The design of a basic heat pump is nevertheless simple and elegant. During the winter, cold air outside a building is blown across a coil of tubing by a fan. The tubing contains a chemical fluid, called a refrigerant, that boils at a very low temperature. The vaporized refrigerant flows through a compressor and exits at high pressure and temperature. Another fan inside the building blows air over the hot refrigerant, whereupon it condenses into the liquid phase, heating the indoor air in the process. The heated air circulates through the building, providing comfort to occupants. Low temperature energy in outdoor air is thus raised, or pumped, to higher temperature with the help of electricity that drives the compressor. Heat pumps are a very robust piece of technology. What's even better is that they work in reverse during the summer, taking heat out of the indoor air and pumping it outdoors. Just flip a switch and you have an air conditioner.

It is important, however, to clarify the concept of efficiency when applied to heat pumps. An increasing number of reports make claims similar to this one: "an air-source heat pump can deliver up to three times more heat energy to a home than the electrical energy it consumes" (US DOE, 2023a). This claim is a reference to the heat pump's coefficient of performance, equal to the amount of heat energy supplied to the building divided by the amount of electric energy used to run the compressor. The 2023 federal standard for minimum heat pump "efficiency" mandates coefficient of performance values between 2.05 and 3.4 depending on the size of the unit (Johnson Controls, 2023; US DOE, 2023b). A heat pump coefficient of performance is not really a direct measure of energy efficiency, but rather an energy conversion factor. One must consider the entire energy value chain, beginning with the production of electricity at a power plant. Some form of primary energy, either fuel (natural gas, coal, uranium), solar radiation, or mechanical energy (wind power) is converted to electricity at a substantial loss. That is to say, the process of converting sunlight or wind or heat to electricity is far less than 100 percent efficient (entropy strikes again!). For example, thermal power plants typically lose 56 percent to 67 percent of their heat energy in conversion to electricity (Kirk, 2022). If we calculate the reciprocals of these energy conversions, the value of $1/(1 - 0.56)$ turns out to be 2.27, and the value of $1/(1 - 0.67)$ is 3.03. This emergence of a ratio between 2.05 and 3.4 is not entirely a coincidence. A heat pump is essentially designed to reverse the prior conversion process that occurs at a power plant by turning electricity back into heat. But heat pumps, like all machines, are never thermodynamically reversible—they too will suffer further operational losses. Think of it like going on a trip to France: you sell dollars to buy euros, paying a transaction fee. At the end of your trip, you sell your leftover euros and buy dollars, again paying a transaction fee. You lose coming and going, just like in Las Vegas.

The upshot is that, although the coefficient of performance quoted for a heat pump might seem impressive on its face, it does not imply creation of free energy. Rather, the operation of a heat pump is merely the second stage of an energy conversion transaction that began at a faraway power plant. There are no cost-free energy transactions, a limitation enforced by the First Law of Thermodynamics (energy cannot be created or destroyed) and the Second Law of Thermodynamics (no energy conversion is 100 percent efficient). Conversion losses are therefore a penalty of home heating electrification that cannot be avoided. The average efficiency of conversion from coal, oil, gas, or uranium generated heat to electricity in 2022 was approximately 31.9 percent, 30.6 percent, 44.1 percent and 32.7 percent, respectively (US EIA, 2023c). The efficiency of conversion from solar energy to electricity using commercially available photovoltaic panels ranges between 17 percent and 20 percent (University of Michigan, 2023b). Conversion efficiencies of wind energy to electricity are typically less than 50 percent (University of Michigan, 2023c). An additional 5 percent of electricity is lost in transmission and distribution (US EIA, 2023d). When you use electricity to run an appliance you incur even more energy losses inside the motor, although electric motors are substantially more efficient than thermal power plants.

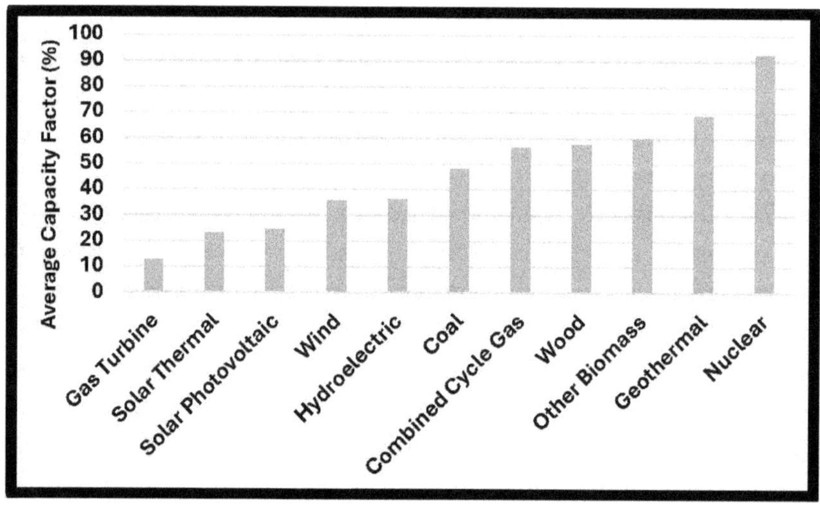

Figure 3-1: Capacity factors of common power plant types (Source: EIA, 2023e)

Power plant efficiency necessarily lies at the heart of all efforts to decarbonize electricity generation and, thereby, the rest of the energy value chain. To accurately assess power plant efficiency, you need to factor in uptime. A nuclear power plant runs almost constantly, going offline only in an emergency or to undergo planned maintenance. Coal and gas plants can also run steadily, although they are down more often than nuclear plants. Solar plants can run only when the sun is shining, and the intensity of solar radiation varies from hour to hour even during the day. A similar energy intermittency logic applies to wind power plants. The fraction of time a power plant spends generating electricity is known as its capacity factor. Capacity factors for conventional nuclear power plants averaged 92.7 percent in 2022; hydroelectric averaged 36.3 percent; coal averaged 48.4 percent; combined cycle natural gas (using heat from burning gas to both drive a primary gas turbine and to boil water and drive a secondary steam turbine) averaged 56.6 percent; solar photovoltaic averaged 24.4

percent; wind averaged 35.9 percent (US EIA, 2023e). The disparity in uptime availability of all common power plant types is illustrated in Figure 3-1.

If you need to guarantee a certain amount of electric power generation, you have to account for both efficiency and capacity factor. This means that a low efficiency, low capacity factor plant must be made larger than it would otherwise need to be to supply a given electrical load. Similarly, to balance the intermittent operation of a solar or wind power plant or to provide emergency capacity during periods of peak power demand, additional generation units—often in the form of gas turbine plants—are built to run only when needed. For this reason, their capacity factors have historically been quite low (EIA, 2023f). Almost any type of power plant can be run at less than its maximum capacity ("spinning reserves") or be temporarily disconnected from the grid ("offline reserves") when demand is low. These practices, while necessary to ensure reliability of service, reduce capacity factors below their maximum achievable levels. Moreover, increasing amounts of wind and solar on the grid, along with different kinds of power consumption such as large backup battery banks and distributed generation (primarily rooftop solar) have created new reliability challenges for utilities and a need for regulators to revise operating reserve requirements (Electricity Advisory Committee, 2019).

EXERGY AND THERMODYNAMIC SCARCITY

As we have discussed, thermodynamics places limits on the strategies we can use to optimize energy flows. Dense energy sources are highly desirable because they contain a large amount of energy in a small volume or mass. Another way of saying this is that dense natural resources are highly concentrated. Concentration of resources does not arise instantaneously. Rather, it is a result of slow, energy intensive chemical and geologic processes. We previously looked at the example of coal, which is formed under conditions of high pressure and temperature deep below the Earth's surface as decaying plant matter is steadily buried by sedimentary rocks. A similar story can be told for oil, natural gas, or any other mineral. Metals are concentrated at specific places over time as minerals are shifted, melted, precipitated, and transformed by geochemical reactions. The energy to do all this concentrating of resources ultimately comes from the sun and from gravitational and electrical forces that bind matter together. When we drill for oil or gas, or mine solid mineral deposits, we are spending some of that stored energy. The elements that comprise these resources do not disappear when we burn fuels or refine ore into pure metals. They are transformed into the products of chemical reactions: hydrocarbons and oxygen are burned and become carbon dioxide; hematite and magnetite are melted in a blast furnace with coke and become steel and carbon dioxide; copper-containing ores are crushed, refined, and separated into pure metals having economic value and waste materials. Pure metals are blended into alloys with precise desired mechanical or electrical properties such as stainless steels, brass, or bronze. All matter has been conserved, down to the tiniest fraction of a gram, but the processes through which it has been transformed are, as always, irreversible.

Once elements have been dispersed and combined into innumerable consumer products and fuels have been consumed, the original raw material components can only—sometimes—be recovered at a significant cost and energy penalty. To regenerate oil from a cloud of carbon dioxide and water, or to recover pure metals from discarded electronic products is technically difficult and will inevitably incur huge losses when compared to the production of virgin materials from concentrated mineral deposits. The celebrated Defense Advanced Research Projects Agency (2022) is hard at work on the problem of metal recovery from electronic waste. Learning can improve process efficiency, and economies of scale can spread fixed costs among large quantities of product units, but it is unrealistic to assume that materials recycled from complex, manufactured chemical compounds will ever be as cheap as those obtained from early, easy-to-mine deposits of minerals.

A key limitation lies in the difference between physical and chemical changes to a material. Physical changes do not alter the chemical composition of a substance. An example of a physical change is the creation of an alloy from several metal elements, like brass from copper and zinc. The individual metal elements can be recovered fairly readily by melting the brass and separating them by exploiting differences in density and boiling point, although the process is still rather complicated (Maleki et al., 2022). Chemical changes, on the other hand, create new, distinct compounds and are difficult, if not impossible, to reverse. Think of trying to recover dry white flour and raw eggs by somehow unbaking a cake; you can't. The chemical recombination of elements into highly engineered, advanced materials presents substantial and fundamental barriers to recycling. Both physical and chemical mixing of materials makes them less available for further use, and engineers ascribe a special meaning to the term availability (hint: it has, yet again, something to do with entropy).

Where a given material exists only in dispersed form rather than as a concentrated deposit it is said to be thermodynamically scarce (Valero et al., 2021). Another way to view thermodynamic scarcity is to recognize that, from a chemical perspective, completely dispersed elements are defined as those that have come into equilibrium with their surroundings at concentrations equivalent to the trace levels at which they may be found at a random location in the Earth's crust. In such a state, engineers would say that elements have low exergy. Exergy is a thermodynamic concept that refers to the maximum amount of useful work that can be obtained by bringing a physical system into equilibrium with its surroundings. The study of exergy in systems is also commonly referred to as availability analysis, because exergy can also be thought of as the amount of usable energy that is available for a particular intended purpose. Accounting for changes in the availability of a system is also a way to quantify the irreversibility of a process. Recall that irreversibility is a measure of the entropy generated during a process, and that high entropy generation is related to low efficiency. Looking at this from the reverse angle, highly irreversible processes can be said to waste availability, or at the very least consume scarce amounts of it. From this standpoint, it is evident that mining techniques that extract tens, hundreds, or in some cases thousands of tons of earth to recover ounces of precious metal are monstrously inefficient and irreversible—even where they are "affordable" in a market economic sense.

Why isn't 100 percent of the energy in any system always available to be usefully employed? Consider heat energy as an example. Heat flows from regions of high temperature to regions of low temperature; hot to cold, and not the other way around (unless you add extra energy to force it to flow backwards, such as when using electricity to run a heat pump). We might say, then, that the energy contained in a hot object is of higher quality than the energy contained in a colder object, because it is easier to move the heat energy in a hot object somewhere else. Ice also contains a lot of energy, in the bonds that hold water molecules in a rigid crystal structure, as well as in the bonds that hold individual atoms together as water molecules. Water molecules are constantly vibrating and rotating and moving around with respect to one another (as are all molecules, all the time, everywhere). Engineers call this kind of molecular motion internal energy, and we measure internal energy as temperature, which is to say, heat. If you want to transfer heat energy from ice to some other object, it must be colder than the ice. That is not generally how we use ice. Instead, we put the ice near a hotter object and allow the hot object to transfer some of its energy to the ice. The ice melts—not because it's losing energy, but because it's *gaining* energy.

One of the reasons that geothermal energy has not played a more prominent role in the clean energy transition is that most of the geothermal heat found in easily accessible locations is of low quality. The best, most available, or most exergetic geothermal energy comes from locations where the rocks are hot enough to boil water underground. The resulting steam rises naturally to the surface and can be tapped at minimal cost to run turbines in an otherwise conventional power plant. This is called a hydrothermal resource and is relatively rare: notable large sites are located in California, Iceland, and Italy, but hydrothermal resources are not going to meaningfully address our global carbon pollution problem. Entrepreneurs are trying to adopt other techniques, using the concepts of exergy and high versus low quality heat, to overcome the limitations of more common geothermal energy sites. Extremely deep, hot, and hard rocks are found nearly everywhere and can be hydraulically fractured at considerable expense, after which cold water is pumped underground and recirculated to the surface once it has absorbed heat from the rocks. The resulting moderately warm water can be cycled through heat exchangers to transfer its lower quality heat to a secondary fluid that boils at a correspondingly lower temperature; that secondary fluid then drives a vapor turbine in the usual manner (DiPippo, 2015).

One of the most interesting recent developments involves a plan to first generate electricity using the highest temperature geothermal water available in a particular location, and then to take advantage of the remaining exergy to drive lower temperature chemical processes or provide low temperature heat for other industrial or agricultural applications. Transporting low quality heat from the outlet of a geothermal plant to the site of an existing industrial facility is expensive, though. It turns out, conveniently, that one method of chemically capturing carbon using catalysts requires heat that is at almost precisely the temperature of the water left over after generating geothermal electricity (Osaka, 2023e). Hélène Pilorgé, a notable researcher in this area, estimates that if all geothermal power plants in the United States were equipped with these carbon scavenging catalysts, 12.8 million metric tons of carbon per year could be captured. This is certainly useful, if only a tiny fraction of what we emit annually.

INDUSTRIAL PROCESS CONSTRAINTS ON DECARBONIZATION

Physical constraints also hinder decarbonization in industries where carbon is essential to the production process, rather than merely contained in fuels. The cement industry is extremely difficult to decarbonize, irrespective of the primary energy source used to manufacture the product. The world consumes thirty billion metric tons of concrete every year (Nature, 2021a). This ubiquitous mixture of Portland cement, sand, gravel, and water is essential to the construction of buildings, roads, bridges, and dams. Portland cement, the glue that binds concrete and gives it strength, is manufactured from an intermediate product called clinker. Clinker is a mixture of minerals that when heated releases large quantities of carbon dioxide as part of its chemical transformation into what will form the bulk of concrete. This heating and chemical transformation must occur, even if carbon free electricity or recycled heat from another chemical process is used instead of fossil fuel to prepare the cement (Hook and Dempsey, 2021). Research is ongoing to reduce the carbon intensity of concrete production, or to replace it with lower emitting substitutes. Even after expected improvements are made, however, at least a 30 percent increase in the cost of construction material is likely to result.

Steel is equally hard to decarbonize, even with wind or solar power to run blast furnaces. Conventional blast furnaces require large amounts of metallurgical coke—produced from coal or crude oil—as a chemical reactant and structural support for transforming raw iron ore into the intermediate alloy that will eventually become steel. Electric arc furnaces can circumvent this step by using recycled steel, rather than iron ore, as an input, but this requires a steady supply of high-quality scrap metal (Hoffmann et al., 2020). Overreliance on recycling would place limitations on the rate of steel production while increasing its cost because supply would be constrained with respect to current levels. Moreover, specialty alloys, such as high strength or corrosion resistant steels, must be blended to carefully controlled compositions that are difficult to obtain from recycled materials alone. To cite another example, plastics and petrochemicals including paints, adhesives, synthetic fibers, dyes, detergents, and pesticides, are fundamentally carbon intensive because they are overwhelmingly manufactured from molecules derived from oil and gas. A few polymers can be manufactured from cellulose, a component of plant fiber, or otherwise decoupled from petroleum feedstocks, but they are exceptions rather than the rule. All our essential building materials are produced in ways that work against efforts to implement a low carbon energy future. We should expect to consume less of each of them if we are to meet climate change mitigation goals.

Even the Scottish whiskey industry is engaged in efforts to develop new, low-carbon heating sources for its distilleries. Wood chips (harvested from trees cultivated for that purpose) and green hydrogen are two leading contenders to replace gas, oil, and traditional peat fires (Booth, 2023). Why not just borrow basic technology from the electric teakettle and dispense with other fuels entirely? Whiskey distillation is an art as much as a science, and the precise rate and temperature of applied heat during the process can have a dramatic effect on the quality of the finished product. Many other processes, including pharmaceuticals production and specialty chemicals manufacturing, have similar precision heating requirements. Electric heating can often be effective, however, if combined with digital control and heat exchange technology.

Consider an illustrative example from Gin Magazine (2018):

> The [distillation] rate is controlled by the degree of heat applied to the charge (i.e. liquid being distilled). Gentler heat promotes a slower rate, while increasing the heat accelerates the process...We experimented with distillation rates when developing our recipe and realised the importance this has. We initially tried a faster rate, but it didn't produce the character we wanted...[Electric] [h]eating elements within the external jacket are controlled by a remote panel, which switches off when the water reaches the required temperature. The temperature is set and carefully controlled by the distiller to maintain the optimal distillation rate. If the water temperature drops by 0.5 degrees centigrade then the heating elements switch on again to maintain a well-controlled and consistent temperature, essential to keep the delicious house-style.

As we move to a more sustainable energy future, a broad cross-section of existing industrial processes will have to be redesigned, perhaps at substantial cost. Maintaining product quality in the face of redesign will often prove to be a challenge.

LIVING WITHIN THE LIMITS

Any sustainable solution to implementing a low carbon energy transition relies more than anything on a cultural attitude shift. Are the richest people in the world willing to consume a lot less? Walk more? Pay more in taxes? Build *and use* the public transit, recreation, and living systems that can reduce carbon emissions and energy intensity to levels that will allow the rest of the world to move closer to their own living standard, all without doing irreparable damage to the environment in other ways? This is a truly radical proposition: asking an advanced civilization to retreat from a position of unrivaled excess and abundance and to exist in a more equitable manner with others. The arguments against it will be fierce and predictable. Don't we work harder, and innovate more, than everyone else and therefore deserve this? Won't others just take advantage of our weakness/generosity and surpass us, making us vulnerable? Answering these questions must be a collective, deliberative, consensus effort. No government can force an entire population to live within its energy and material means. The temptation to assume that yet-to-be-discovered technology will eventually resolve all of these difficulties is seductive. In the next four chapters, we will explore in more detail the physical and economic limitations and tradeoffs that pertain to each of the primary facets of the energy transition.

CHAPTER 4
UNIVERSAL ELECTRIFICATION WILL BE INCONVENIENT

A few years ago, my wife and I were visiting her mother for the holidays and wanted to buy her a new digital TV antenna. We were quickly reminded that her house, built decades ago, was unequipped for modern, three-prong electrical plugs. The old, two-prong plugs and receptacles were eventually recognized as being dangerous, because an accidental short-circuit or power surge could send a large current through a person or a valuable electronic device, causing death or damage. The National Electrical Code (NFPA, 2023) does not require an existing home to upgrade old electrical receptacles, but newly constructed buildings must include them. Additionally, homes with outdated wiring in which old receptacles—but not the wiring behind them— are replaced must clearly mark the new receptacles with a warning that the third grounding wire is not present. The traditional role of building codes has always been to, first and foremost, ensure the safety of occupants. Sadly, the drafting of building codes today is less about increasing safety than an opportunity for equipment manufacturers to create and defend markets for their products. Electrification advocates are consistently among the most prominent of them.

Many people might assume that the committees that oversee and vote on changes to building codes are staffed by unbiased teams of technical experts. Meetings to discuss building code amendments are nominally open to the public. They are, in fact, overwhelmingly attended by representatives of an impressive array of trade associations: natural gas; pools and spas; aluminum siding; wood pellets. If something can be built into or used in or on a residential or commercial building, you can be sure that a lobbyist is assigned to make sure nothing allows an inspector to deem it out of code. Since they ultimately acquire the force of law once adopted by state legislatures, building codes have also acquired a new mission: they are being used by environmental advocates as platforms to launch decarbonization battles (Harvey and Gillis, 2022). The two missions—product endorsement and climate change mitigation—align perfectly for many of the largest home energy service industries. Universal electrification of all things that currently consume fossil fuels, from gas stoves and water heaters to the oil-fired furnaces of New England, is a huge opportunity to reduce carbon emissions as well as an eye-watering bonanza for appliance makers and those who provide the electricity to run them.

Rooftop solar installers, heat pump manufacturers, purveyors of LED lighting, and electric vehicle charging station or home storage battery vendors are among the many people who are eagerly surfing the electrification wave. Many, even most, of these products are useful and do have the potential to reduce overall carbon emissions. Whether or not they should all be made mandatory features of new or existing buildings is open to debate. Replacement of all existing energy consumption with electric alternatives will be expensive. In some cases, the new products will offer superior performance with few downsides; in others, they will require difficult and inconvenient adjustments to consumer behavior.

THE CHALLENGE OF UPGRADING THE GRID

Electrification is undoubtedly an important pillar of the clean energy transition. On the other hand, it is not a cure-all solution to our energy, material scarcity, or pollution dilemmas. The electric grid is a complex web of wires, transformers, circuit breakers, and capacitors that is growing increasingly difficult to stabilize and repair. Electrical transmission wires are also like leaky buckets—only a fraction of the energy used to generate electricity arrives at your home. It is also undeniable that the carbon intensity of the United States (or virtually any other national) electrical generation system has failed to fall to anything close to the level required to prevent the rise of global average temperatures. We should modify the leaky bucket metaphor slightly because electricity cannot be stored in the sense that water can be poured into a container. Electrical energy must be consumed the very moment it is generated. Imbalances between the instantaneous supply and demand for electricity at any point along the grid can cause instability that, if not immediately remedied, can result in a cascading wave or chain of power failures. These failures may, in severe cases, propagate across entire regions.

The frequency and duration of power failures have been increasing (EIA, 2021; MacMillan and Englund, 2021; Brown et al., 2022). Many of these failures are attributed to climate change induced storm activity (Ross et al, 2022). Some of the worst affected areas are those where the local population is least able to cope due to illness, age, poverty, or lack of transportation (Kim, 2023). Solutions to these problems cannot be implemented rapidly or comprehensively, even where the funding and political will exists. Rushing to electrify the entire economy in an infrastructure-deficient environment is destined to create, at a minimum, inconvenience and, at worst, risks to health and safety. Encouraging greater electricity consumption in the full knowledge that it can only be delivered by a distribution system that was designed for substantially less power is irresponsible. Distributed generation, solar panels being the most feasible option today for most people, can certainly ease the burden when sunlight is plentiful, although too much sunshine can be as detrimental to grid stability as too little power generation. Home battery systems may be able to supply a few hours of backup power if homeowners (or landlords) can afford them. Ironically, the revenue that is subsequently lost by the local utility leads to lower investment in the surrounding grid. It may become even less stable as it must continue to serve loads from connected customers and to cope with the "backward" flow of home-generated electricity on low demand days.

City and state governments are especially concerned about finding ways to upgrade and safeguard electrical generation and transmission systems in the face of climate-related extreme weather events while still keeping those systems affordable. Sweeping and long-lasting power outages are extremely dangerous: in the summer a lack of air conditioning can lead to heat stroke; in the winter a lack of heat is equally lethal to the elderly, the sick, and the poor who have their utility service disconnected for nonpayment. These problems are glaringly visible and embarrassing for local leaders. Despite the obvious need to act, proposed solutions are controversial and require compromises and tradeoffs, as we have seen throughout our discussion of the energy transition.

A prime example involves plans by Texas to prevent a repeat of the devastating electric grid failure in February 2021 that followed a severe ice storm. The Texas Legislature proposed an energy insurance program that would create a privileged investment environment for new natural gas generating capacity to serve as a stabilizing power supply. In 2023 this revised plan was implemented as the Texas Energy Fund (PUCT, 2025). Its benefits are difficult to assess. On one hand, builders of competing power plant technologies make a valid point in claiming that the exclusion of all energy sources other than "dispatchable" (which, practically speaking, means natural gas or coal) may harm the market's ability to deliver cost-effective electricity. On the other hand, from a grid perspective, there are only three realistic options for growing meaningful baseload power capacity: natural gas, coal and nuclear (geothermal is simply not available at the necessary scale). Coal is the highest carbon intensity form of electricity generation, and the mitigation of coal emissions would require a massive acceleration of carbon capture and sequestration, a technology we will discuss in the next chapter. Nuclear power plant permitting is so difficult that it makes that option all but irrelevant (although the recent surge in demand for data center power is sparking fresh interest in nuclear energy, as is a 2024 agreement by 31 countries to triple nuclear capacity by 2050 [Crownhart, 2025]). Utility scale storage, in the form of huge batteries or other commercially available technologies to balance fluctuations in wind and solar generation, is both expensive and unproven at a statewide scale. The Texas Energy Fund is currently considering a Backup Power Package program that would provide grants and loans to build smaller systems to provide multi-day backup power at individual facilities.

Thus, we are currently left with natural gas (Lee, 2023) as the most logical near-term grid stabilization medium. The Texas plan is by no means perfect. Questions surrounding a guaranteed financial rate of return for inactive reserve plants and the wisdom of having so much infrastructure sitting idle so much of the time have still not been definitively answered. A potential saving grace could be designing all new gas turbine generating plants to eventually burn green hydrogen in addition to methane (natural gas). We will see in a future chapter that hydrogen—the lightest element in the universe—turns out to be an excellent way to store electrical energy, and for a longer time and in larger quantities than even the best batteries could hope for. Unfortunately, hydrogen is so chemically reactive that it does not exist in pure form on Earth in meaningful quantities. Therefore, it must be manufactured at considerable expense. It is also highly flammable, explosive, and can damage common pipeline materials.

Electricity supply will also become more difficult to expand as the clean energy transition grows in scale because no one wants to live next to a power plant, even a renewable one. Texas lawmakers have proposed plans to limit growth of the wind and solar industries statewide by increasing siting and permitting requirements (Phillips, 2023). Pressure from rural landowners concerned about visual pollution and property devaluation are driving these moves. This comes at a time when federal tax credits for renewable energy technologies are substantially more generous than usual thanks to the Inflation Reduction Act (although the Trump Administration has threatened to wholly or partially repeal the Act [Fujii-Rajani and Patnaik, 2025]). Natural gas electricity generation is, however, seen as a more favorable alternative by both proponents of the proposed regulations and the new administration. A tradeoff could arise in terms of overall economic productivity. Texas leads the nation in wind power generation and has been sharply increasing its share of solar photovoltaic generation as well. Many of those projects are built in areas of the state that rarely see the kinds of high value investment capital that flows to cities like Austin,

Dallas, or Houston. Property tax revenue forms the basis of school district funding across the state, and job growth of any kind can be hard to come by in far West Texas.

A potential workaround, once again, is to combine more renewable electricity planning with hydrogen. Hydrogen's potential as a substitute for natural gas fuel in electricity generation is huge. Gas turbine power plants built today with an eye toward future consumption of hydrogen are less likely to become stranded assets before the end of their useful lives. Natural gas infrastructure that is already sited and permitted can be retrofitted in many cases to carry hydrogen, albeit possibly at lower pressures. These retrofits can include pipelines, liquefied natural gas (LNG) regasification terminals, storage facilities, and delivery systems. The conversion of equipment from natural gas to hydrogen service is not cheap, but estimates made by Gas Infrastructure Europe indicate that it would be less expensive than building a dedicated hydrogen production and transportation network from scratch (Hall, 2023). A consortium of thirty-two European infrastructure operators have proposed a network comprising 60 percent repurposed natural gas pipeline and 40 percent purpose-built hydrogen pipeline to span 53,000 km by 2040 at a projected cost of between €80 and €143 billion ($90.8 and $162.4 billion) (EHB, 2023). The ability to sidestep land use conflicts by leveraging existing right of way corridors may well prove to be the most decisive benefit in many locations (Findlay, 2020).

A strategy that acknowledges the need to prepare for electrification by pre-investing in infrastructure and deferring electrical loads where feasible can accelerate the delivery of clean power, eliminate more carbon emissions, and enhance grid reliability. Amazon has resisted attempts by the Oregon legislature to mandate 100 percent carbon free power for data center cooling by 2040 (O'Donovan, 2023). The company has a point. Permitting and grid interconnection delays for new clean power projects are placing the most ambitious corporate climate change mitigation goals at risk while giving political cover to those who defer action (Bloomberg NEF, 2023a). In contrast, Texas became the nation's largest producer of wind energy by wisely putting the horse before the cart. The Competitive Renewable Energy Zone

(CREZ) program, begun in 2005 and completed in 2014, ensured that transmission infrastructure was in place first to smoothly deliver future clean electricity later from the remote rural areas where it was generated to the urban centers where demand was highest. The certainty that CREZ offered investors made Texas wind projects much easier to finance (Jankovska and Cohn, 2020).

There is no doubt room to streamline the permitting and interconnection processes. We can also limit competition among new clean power projects by simultaneously leveraging other low carbon energy sources. Low emission "e-fuels" synthesized from captured carbon and green hydrogen have the potential to displace a great deal of future electricity demand. The Center on Global Energy Policy at Columbia University (Friedmann et al., 2019) discusses this strategy in the context of heat-intensive heavy industry. Where deployed for the heating of homes and medical facilities, synthetic natural gas, blended into our extensive existing gas pipeline network, can not only reduce electrical load requirements, but also help provide crucial energy reliability guarantees during power outages. As you might imagine, none of the options Columbia studied are easy or cheap to implement.

Recall that global electricity infrastructure spending needed to support decarbonization will reach at least $21 trillion (Bloomberg NEF, 2023b). Reuters reports that the U.S. share alone of this total may exceed $2 trillion (McLaughlin, 2023). For perspective, the Inflation Reduction Act contains $500 billion of clean energy investments, not all of which are related to electric infrastructure (McKinsey & Co., 2023a). Meanwhile, the United States spent $8 trillion fighting in Iraq and Afghanistan over a twenty-year period (Brown University, 2023). There are only three feasible means of paying the costs of modern, universal electrification: increased taxation, higher government debt, or ratepayer charges. The public will pay in every case, and the only question is how and on what basis. There is an argument to be made that a consumption-based rate charge tied directly to power usage is the most equitable. But we must then be prepared to risk causing a great deal of economic pain for large portions of the population, many of whom may then be forced into energy poverty. That will slow adoption of electric vehicles, heat pumps, and reduce the general tax receipts needed to fund electrical infrastructure projects.

THE ENERGY-WATER-FOOD NEXUS AND ELECTRIFICATION

Water, being essential for life support, is an incredibly sensitive political topic. Citizens greet water price increases or supply curtailments with near-universal outrage, keeping taxes on water consumption low. One of the most visible and immediate effects of a changing climate has been the alteration of regional rainfall patterns. The change is leaving large swaths of farmland and many cities in a state of unexpected and persistent drought. Rahat et al. (2024) note that California in particular faces a "dual challenge of floods and droughts due to elevated temperatures, *variable precipitation patterns*, and increased evapotranspiration, contributing to the risk of wildfires and ecological disruptions" [emphasis added]. It is impossible to separate the effects of energy use on the availability of water, or vice versa. Thus, the need to safeguard and transport water supplies is further complicating moves to secure new sources of energy. The energy-water nexus is a well-understood phenomenon. The exploitation of water resources requires large amounts of energy irrespective of its type. All thermal power plants, whether fueled by coal, natural gas, or uranium, continuously boil water to make the steam that drives electric generators. The manufacture of any chemical product, including solar panels and the components of wind turbines, requires water. Water pumps use energy, and hydroelectric or geothermal power plants, by their very definition, cannot operate without access to large water sources. Roughly 40 percent of U.S. electricity is generated by burning natural gas (US EPA, 2023a), and a single hydraulically fractured gas well can consume as much as sixteen million gallons of water (USGS, 2023a). The price of electricity is (usually) a fairly reliable market signal that reflects the supply of and demand for gas. It is less clear whether water is adequately factored into electricity—or any other—cost associated with its usage (D'Odorico et al., 2020).

Energy and water are also connected by their joint roles in food production. Growing organisms of all kinds require large amounts of energy and water. Crop yields sufficient to feed the world's current population of nearly eight billion people cannot be sustained without supplementary irrigation, which requires energy for canal construction and pumping. One possible way to reduce agricultural energy consumption and carbon emissions simultaneously without expensive new infrastructure is to offer subsidies that encourage farmers to switch to drought-tolerant and climate-flexible crops. Another obvious, but incredibly controversial, idea is encouraging people to alter their diets in ways that reduce fertilizer production and water use. What we eat and how we produce our food must be a part of the conversation about any serious decarbonization plan. Agricultural practices are responsible for as much as 29 percent of global carbon emissions (World Bank, 2023a). Meanwhile, approximately 33 percent of all food produced worldwide is wasted. Our food production system is inefficient and contributes to climate change patterns that threaten to increase hunger even more. For many years, a small group of farmers, bakers, and chefs have worked to cultivate and prepare high protein foods made from insects, because raising livestock—especially cattle—is a land and water intensive operation that generates substantial amounts of fugitive methane emissions. That's right: they want you to eat bugs instead of steaks.

Cultivation of crickets, grasshoppers, and similar insects can be a far less resource intensive practice than grazing conventional livestock. Will it ever be accepted in a country like the United States? Switzerland's approach is to start with young children who have fewer preconceived notions about what is normal, acceptable, or proper (Dalton, 2023b). Although Switzerland was the first European country to approve the commercial sale of insect-based foods (McKenna, 2017), consumer attitudes there are still mixed (Ibrahim, 2022). Many people all over the world eat insect protein, with acceptance of the practice varying quite a bit depending on culture, marketing, and overall food availability (Guiné et al., 2024; Alhujaili et al., 2023). Insects are still a decidedly niche market in the United States, where "regular" meat consumption has become a culture war issue: Conservative politicians responded with outrage to false rumors that former president Biden had been planning to ration red meat (Beaumont and McFetridge, 2021).

This isn't the first time a controversial discussion about food choice has occurred. The dairy industry fights against labeling plant-based beverages as milk. Skeptics of laboratory raised meat question its safety, if not its flavor. A fascinating story about the past and future of U.S. livestock production is Jon Mooallem's highly entertaining book *American Hippopotamus*. You may be surprised to learn that Congress very nearly voted to import hippos from Africa at the turn of the twentieth century, but the cattle industry convinced the lawmakers that beef was "a [more] normal meat to eat" (Mooallem, 2013). During both world wars efforts were made by the American civilian population to abstain from eating meat for the benefit of soldiers serving overseas. The government encouraged civilians to substitute kidneys, hearts, and other organ meats during World War II (Romm, 2014). Conservation of food during wartime was also seen as a patriotic duty during World War I as reported by Romm (2014):

> "Meats and fats are just as much munitions in this war as are tanks and aeroplanes," wrote [Herbert] Hoover, who led the U.S. Food Administration during World War I (and pioneered the slogan 'Food will win the war,' as well as the first Meatless Mondays). "The problem will loom larger and larger in the United States as the war goes on … Ships are too scarce to carry much of such supplies from the Southern Hemisphere; our farms are short of labor to care for livestock; and on top of it all we must furnish supplies to the British and the Russians."
>
> "We should not wait for official rationing to begin to conserve," he continued. "The same spirit in the household that we had in the last war can solve this problem."
>
> Hoover knew of what he spoke. Just two months later, meat would join butter and cheese as a rationed food item, as growing quantities of beef and pork were shipped overseas to feed American and Allied troops.

Food conservation can be nearly as impactful to climate change mitigation as increasing energy efficiency. It also flies in the face of some of our most deeply held cultural traditions and personal habits. As evidenced by the reflexively negative political response to a mere rumor of an idea that meat might be rationed in today's America, a repeat of the "Meatless Mondays" policy does not appear to be on the horizon.

Water problems are also often electricity problems. Few American water conflicts are more visible and contentious than the fight brewing over future allocation of the Colorado River. Since 1922, seven U.S. states (Arizona, California, Colorado, Nevada, New Mexico, Utah, and Wyoming) have shared water from the Colorado River under a compact mediated and approved by the federal government (USBR, 2023a). Mexico was also belatedly added as a treaty member. Persistent drought and population growth have combined to create intertwined challenges that threaten to upend life in the American West. Farms in what would otherwise be parched lands in Southwestern states depend on river water to grow America's produce. And it isn't just that we need the food, but also that we insist on eating whatever we want whenever we want to, whether it is "in season" or not. Without the Colorado River compact, there would be a much smaller selection at your local grocery store or favorite restaurant during much of the year. Meanwhile, the same river water is expected to flow through giant hydroelectric dams to power western cities.

The Imperial Valley of California on the western bank of the river, and the Yuma, Arizona, area on the eastern bank, grow as many as 90 percent of the vegetables available to Americans between November and March each year, when the rest of the country is too cold to contribute to the harvest (Satran, 2023). Of course, we also import a lot of food—especially our favorite fruits and vegetables that we want all year long, but don't grow everywhere all the time (USDA, 2023a). The American melting pot is a more apt metaphor than we sometimes realize, and our collective culture has benefited from the amalgamation of cuisines. A prime example of this culinary pollination is the introduction of plants and animals—apples, bananas, grapes, beef, and chickens—by Spanish and Portuguese explorers to the Western Hemisphere beginning in the sixteenth century: perhaps no other immigrant group has ever done so much to influence the eating habits of so many (Baer-Sinnott, 2023). Unfortunately, not all these ingredients grow equally well at all times or in all places. Climate change, electricity demand, and water scarcity are threatening to force us to learn to live without many of our common foods. Alternatively, we will have to develop expensive new sources of electricity to adjust for water unavailable from the Colorado and instead move more water from other stressed lakes, rivers, and aquifers. If not, we will have to rely more on food imports, or we will have to use less water for domestic consumption, golf courses, and manufacturing.

Seasonally adjusted eating need not be a bad prospect. Before the advent of modern supermarkets and refrigerated transport, farmers' markets often supplied most of our produce (Taggart, 2023). These items were by definition and by necessity locally grown and seasonally available. Rich cultural traditions often revolve around memories and practices associated with local food cultivation and preparation, as related beautifully by Albert (2018):

> Storytelling is an important part of Cherokee culture. Many of our stories not only contain lessons about how we nourish our bodies but also about where we come from and how we should treat one another. When you hold a strawberry in your hand or pull the husk from an ear of corn, knowing the stories that accompany these foods transforms them. They become more than an ingredient to measure and cook with for a specified amount of time. That strawberry becomes more than just a berry. That ear of corn is so much more than "just an ear of corn." It is an ancestor, it is our mother; a reminder of who we are, what we've been through, and why we must continue to survive. Every meal has the potential to be a small ceremony, a direct link between our ancestors before us and the future of our people.

Substitution of one vegetable for another seems like an achievable objective. I can imagine a fair amount of complaining, but I would not expect people to riot over heirloom tomatoes. In contrast, a permanent transition to a meatless, or even a meat-minimal diet, will be impossible to enforce in the absence of resource constraints that make livestock production uneconomic for all or most producers. Most people really like to eat meat. You'd have an easier time banning coffee.

The electricity challenge associated with the Colorado River is just as serious as the food challenge. Energy policies in southwestern cities have been predicated for a century on the assumption that large amounts of hydroelectric power will always be available from dams. These megastructures, many of them legacies of the New Deal, are at risk of ceasing to operate entirely, because "[t]he period from 2000 through 2022 is the driest twenty-three-year period in more than a century and one of the driest periods in the last twelve hundred years. This has resulted in historically low reservoir levels at Lake Powell and Lake Mead" (USBR, 2023b). Lake Powell is, as of the time of this writing, only thirty feet above the minimum level needed for Glen Canyon Dam to produce electricity. Although extraordinarily large winter snowfalls in 2023 have delayed the day of reckoning, without lasting and sweeping changes to water use and allocation that day will come. Some 4.5 million people depend on the power from Glen Canyon; cheap, carbon-free power that will have to be replaced (Partlow, 2023b).

The non-adjustable water amounts allocated to each state under the treaty are a huge flaw in the design of the Colorado River compact. The original 1922 agreement reserved 7.5 million acre-ft of water for the "upper division" states (Colorado, New Mexico, Utah, and Wyoming), and another 7.5 million acre-ft of water for the "lower division" states (Arizona, California, and Nevada). In 1944 Mexico was allotted 1.5 million acre-ft, bringing the fixed annual total amount of water distributed from the river to 16 million acre-ft per year. At that time, the maximum annual flow of water from the river was estimated to be 18 million acre-ft (Koebele and Robinson, 2021). Engineers often speak in terms of safety factors, margins above the design values of key variables that are added to allow for inaccuracies in their calculations or unexpected changes in the ways their products are used. The 1944 Colorado River allocation represented a safety factor of 12.5 percent, which seemed reasonable.

Unfortunately, the assumptions underlying the compact turned out to be poor ones. When he was commerce secretary, Herbert Hoover, the federal government's representative during negotiations, predicted that the 1915 population of the Colorado River basin (457,000) would eventually quadruple. That population currently stands at forty million, more than twenty times the 1915 level. It was a terrible prediction, in hindsight (Osborne, 2023). An even more troublesome error involved prediction of the river's maximum water flow, estimated during an unusually wet period. The actual annual flow since 2000 has been approximately 12 million acre-ft per year (Osborne, 2023), 25 percent less than the allocated flow. There is no mechanism for readjusting the allocation of Colorado River water usage based on reductions in flow or population increases. The treaty is the treaty, and no one really wants to renegotiate it. The livelihoods of everyone from family farmers to Las Vegas casino dealers are at stake. As of the time of this writing, the states involved in the compact are discussing compromises, but any changes are unlikely to represent comprehensive reform.

Similar stories of water conflict are unfolding all over the world. They span the Nile River in Africa, the Indus River where India, Pakistan and China meet, and the Mekong River basin in Southeast Asia. Whose access to water is most vital: those who wish to farm; those who wish to produce hydroelectricity; urban areas; rural areas; upstream versus downstream users? In the short term, solving the water problem is almost guaranteed to result in a larger amount of coal- or gas-fired electricity generation in lieu of hydroelectricity, especially coal in the developing world. Relying on coal-fired power capacity growth at the expense of hydropower would make universal electrification a far less compelling climate salvation narrative.

ELECTRIC VEHICLES

The National Renewable Energy Laboratory (2020) conducted a much cited RE Futures study that considered scenarios for changing the electricity source/ energy supply mix over the coming decades. Under a high electrification scenario, in which 86 percent of vehicles are electric by 2050, generation capacity would have to double to support a 67 percent increase in power demand. The authors concluded that a more likely scenario would result in only 40 percent of vehicles being electric by 2050, which of course means that 60 percent will use other—almost certainly hydrocarbon—fuels. The mathematical and economic models used in this study tested maximum renewable electricity generation levels between 30 percent and 90 percent, with the most plausible level at 80 percent. The 80 percent threshold is significant, because it corroborates other challenges involved with maintaining smoothly operating wholesale power markets at near-total shares of intermittent (wind and solar) generation. These power market challenges are likely to reinforce upper limits on the extent to which we can replace conventional baseload fuels such as natural gas or nuclear fission. We will return to the topic of energy market failures in a later chapter.

Model	DC Fast Charging Speed (miles of range per hour of charging)	Minutes required to acquire 406.4 miles of driving range at DC fast charging speed
Porsche Taycan Plus	650	38
Kia EV6 Long Range 2WD	650	38
Mercedes EQS 580 4MATIC	490	50
Tesla Model Y Long Range Dual Motor	370	66
Hyundai IONIQ 5 Long Range 2WD	580	42
Audi e-tron GT Quattro	640	38
Polestar 2 Long Range Single Motor	340	72
BMW i4 M50	370	66

Table 4-1 Best electric vehicle charging speeds available in the United States as of 2022 (Source: Electrek)

Financing and market structure tradeoffs are not the only practical constraints on vehicle electrification. Can we ever expect to have the same range and recharging speed we have become accustomed to with liquid fuels? Technology is improving, we are assured, but consider the following list of some of the fastest charging electric vehicles available for sale in the United States in 2022 (Doll, 2022). Table 4-1 shows their fast-charging rates—rates attainable only with the most advanced (direct current [DC], also known as Level 3) vehicle chargers, which are a small minority. These rates are also shown as the equivalent time needed to acquire the same range (406.4 miles) represented by a full (sixteen gallon) tank of gasoline at the record-setting 2021 average U.S. fuel economy of 25.4 miles per gallon (US EPA, 2021). One can easily refill an automobile gasoline tank of that size in five minutes.

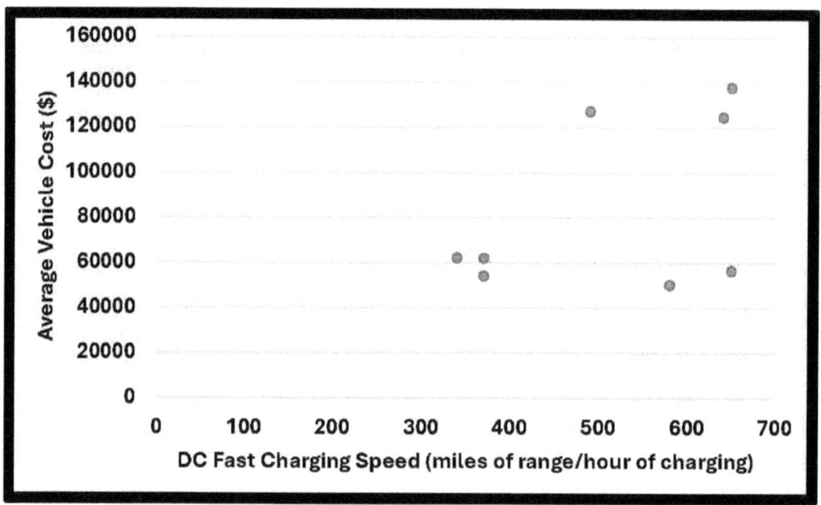

Figure 4-1. Electric vehicle cost as a function of DC fast charging speed (Source: CarsDirect, 2023)

Figure 4-1 illustrates the relationship between electric vehicle charging speed and upfront cost, the latter averaged across available trim levels (charging speed is assumed to be constant for a given model). The good news is that the best charging performance is not highly correlated to vehicle cost—manufacturers can and do offer less expensive models with relatively rapid charging speeds. On the other hand, none of these are what could reasonably be called economy cars: the all-in price of a modestly equipped 2019 Toyota Corolla was approximately $22,000 (Clarke, 2022a). We should expect to invest a minimum of $50,000 in an electric vehicle, when the median U.S. household income in 2021 was $70,784 (U.S. Census Bureau, 2022).

In one of these typically two-income households, a pair of electric vehicles for daily commuting will require the equivalent of seventeen entire months of income—at a bare minimum—and only if the entire sum is paid in cash. Financing can add several thousand dollars in interest payments to that figure (Clarke, 2022b). Chinese-made electric vehicles are often less expensive and are beginning to gain market share abroad. However, governments in Europe and North America are wary of allowing themselves to become overly dependent on foreign batteries, minerals, and other strategically important technologies. Trade restrictions—tariffs and domestic material content quotas—currently prevent this dependence, which raises consumer prices for imported vehicles.

Leaving aside for a moment the large increase in electric vehicle cost compared to conventional cars, the order of magnitude increases in refueling times will drastically affect our travel and commuting habits. If you've ever taken a long road trip in the United States, you'll be familiar with some of the many small, isolated towns along our vast stretches of interstate highway. Travelers pull in for gas, drinks, and maybe a fast meal and quickly move on. Waiting for a pump at the gas station is a minor annoyance that might cost you a few minutes on a busy holiday weekend. When refueling (recharging) lasts forty, sixty, or more minutes per electric vehicle, waiting becomes a serious problem. A line of just three or four cars ahead of you could cost you half a day's travel time. One proposed solution is to make charging available at motels. But unless the chargers are continually kept in excellent operational condition and installed in numbers adequate to serve a full complement of parked vehicles, many travelers won't be able to access them. Put simply, traveling across the country by car will demand a lot more patience and flexibility. Increasing the range obtainable from a single charge, or consumer preferences for vehicles with very large batteries (think electric trucks and SUVs), will only exacerbate the problem: bigger batteries take longer to recharge, all other things equal.

On a more positive note, a return to slower paced travel also suggests that demand for more roadside services, from lodging and food to auto repair, retail, and healthcare, could revitalize rural areas. Combined with greater opportunities for remote work, a modest amount of migration away from cities could alleviate overcrowding, traffic congestion, housing price inflation, rising crime, and air pollution. An influx of population into rural counties might also eventually help reduce some of the mistrust, political polarization, and lack of ecological awareness that stem from the psychological and physical distance that urbanization has created among us over the past several generations. More people might become reacquainted with where exactly their food and water come from. We might get to know our neighbors better, even if there are fewer of them. We might even finally decide that a national, high-speed passenger rail network is a good idea.

In contrast, the suburban commuter lifestyle will likely not survive a transition to completely electrified transportation. The hectic rush from home, to office, to after school pickup, to soccer practice, to the grocery store will eat up daily range that cannot be instantly replaced. Commuter train service will not support personalized errand schedules like private vehicles can. Some people will work remotely, but many cannot. Those who provide face-to-face service will move closer to city centers. Property valuations will shift, and families who find themselves downtown for the first time will have to adjust to having less personal space. Employers who demand face time in the office will ultimately be subjected to the same physical limitations of electric vehicles that their employees encounter.

All of this is to say that Americans may start spending more time in one place than we've become accustomed to at any time since at least the 1960s. Towns along what is now Interstate 25 in New Mexico were spaced seventy miles apart to allow for the resting of horses traveling El Camino Real, the Western Hemisphere's first transcontinental superhighway. This colonial road brought Spanish settlers and supplies from Mexico City to Santa Fe between 1603 and 1821 (US BLM, 2023). People could not, and did not, travel hundreds of miles per day. They took advantage of the hospitality in towns along the way more than we typically do on our modern road trips. In the days of horse travel, trips were less frequent, and stays at destinations were longer. The arrival of the railroad ended the era of the Camino Real. A return to rail travel may be the best available, if imperfect, alternative to the challenges of high cost, short range, slow-to-charge private electric automobiles for intercity transport.

Hydrogen has struggled to gain traction as an alternative motor fuel. The market, instead, appears to have largely settled on plug-in electric vehicles, at least in terms of light- duty passenger cars and trucks. Hydrogen is better positioned in the near to medium term to become a low carbon substitute for natural gas in stationary applications where it can be delivered easily by pipeline. Electric vehicle charging infrastructure, in addition to being sparse at the time of this writing, is also complicated. There is no good reason for this to be the case, and it is one of the major impediments to mass consumer adoption of electric vehicles (Osaka, 2023a). There are at least five charging plug interface standards currently in use (Doll, 2021), along with several proprietary plugs that can be adapted to fit a standard household 110 V outlet. But, as mentioned previously, a potentially bigger problem is charging speed. The larger the battery, the longer the vehicle's range, all other things equal. But, again, this range comes at the price of longer recharging time. Ubiquitous fast Level 3 charging is the industry's goal, but, using the slower, more widely available Level 2 (an alternating current [AC] technical standard), charging a Ford Mustang Mach E can take over eleven hours (Doll, 2022). If most people are expected to charge at Level 2 speeds at home, constraints on intercity travel or long commutes coupled with after-work errands and leisure activities become even more severe. And, for comparison, Level 1 (also AC) charging is the even slower speed at which you would have to recharge your vehicle by plugging it into a regular household wall socket. Daily reliance on Level 1 charging alone would be totally unacceptable for anyone who travels more than one charge to/from his/her office and/or must run errands.

The federal government understands the importance of a unified EV charging standard. The Biden Administration promoted this goal by requiring a single standard for all federally funded charging stations (White House, 2023a). That is good, but it will not necessarily apply to privately funded stations. In this arena, there are several competing visions for charging station networks. Automakers such as Tesla are building their own networks as they see fit; retailers such as Walmart plan to build chargers in their store parking lots to attract customers; individual cities are considering their own local networks; and the organizations that write building codes are hoping to mandate that charging capacity be wired into every new structure—residential or commercial.

Tesla is currently winning the technical argument, with major automakers General Motors and Ford adopting that automaker's North American Charging Standard (NACS) as of 2023. Volkswagen is considering a switch from the Combined Charging Standard (CCS) to the Tesla standard as well (Bushard, 2023). Smaller manufacturers, including Rivian, Polestar, and Hyundai are also either committed to or seriously considering NACS (Shaw, 2023). These moves could make U.S. government intervention less relevant because no additional standards consolidation work would need to be done. Incredibly, however, in April 2024 Tesla laid off nearly the entire staff of its Supercharger business unit and announced plans to defer expansion of its vehicle charging network (Hiller et al., 2024)! The company subsequently backtracked and rehired some of its charger development staff. Nevertheless, we still find ourselves in need of coordination among competing efforts to build physical charging station networks, regardless of the standard. We need to install the charging technology we need where we need it, and to ensure that consumers can count on a predictable level of charge at every location so they can plan long trips. Failproof reliability of individual chargers is especially critical, since a motorist will be less able to carry an extra battery for emergency recharging compared to toting an extra ten gallon can of gasoline. Portable solar panels may deliver very small amounts of charge over many hours, and roadside assistance providers may roam the highways with charging trucks, but such workarounds will not be welcome or feasible on a routine basis.

Serious questions about the pace of electric vehicle rollouts in the United States are beginning to arise (Colias, 2023; Davis, 2023; St. John and Naughton, 2023a; St. John and Naughton, 2023b). As we have noted, vehicle cost and charging performance are the most immediate consumer concerns. Another factor to consider: This buyer hesitancy is occurring as profound shortages of many scarce minerals loom in the next few decades. Lack of minerals such as lithium, cobalt, and nickel will further depress demand for electric cars, because batteries that require these materials are the single largest cost component in a plug-in vehicle. Supply chain complications will frustrate the plans of electrification advocates in other ways as well, as we will see in the following discussion of demand response.

CONSERVING ELECTRICITY BY CONTROLLING DEMAND

The electric grid is aging and fragile. One often discussed replacement plan involves shifting from reliance on a single, massive grid to a decentralized system of rooftop or small community solar plants augmented by battery backup (Coren, 2023b). Since today's home batteries are large and expensive (an installation cost of $15,000 to acquire a day's worth of reserve power), electric vehicle batteries with two-way power flow are suggested as a workaround. The idea is that your car will power your home when you're not driving. Such a multi-purpose system may appeal to relatively affluent homeowners who do not rely on their electric vehicles for long commutes or frequent time-sensitive errands (childcare, medical visits) and can afford the most sophisticated models (and who can tolerate occasionally walking into the garage to find the car battery dead). Nevertheless, a tight coupling of transportation and residential energy supply is problematic in several ways. First, apartment dwellers and homeowners in areas where solar power is inefficient will be poorly served. The more of these systems are installed, the less ratepayer revenue will be available to electric utilities for maintaining the public grid. As such, we risk creating a downward spiral of decreasing electric service reliability for millions of Americans. Second, vehicles serving the grid for load balancing purposes that fail to meet all of their owners' daily travel needs could lead to substantial losses in overall economic productivity, or to dangerous fluctuations in home energy availability if vehicles are disconnected. If privately owned vehicle batteries become a substitute for public grid investment, regional energy resilience will degrade. Third, supplies of critical minerals for battery production, as well as the processing and manufacturing capacity for them, are concentrated in only a handful of locations worldwide. Adapting the role of vehicle batteries to meet essential home electricity needs presents a potential risk to energy security for most nations.

Many see sodium as a much more available and less expensive alternative to lithium for batteries. The periodic table of the elements is so named because the same types of chemical properties tend to reappear periodically in atoms having the same outermost distributions of electrons. Sodium, like lithium, has a single electron in its outermost shell, and this electron configuration is one of the key reasons why lithium is so prized as a battery electrode material. Sodium is significantly heavier than lithium and the energy density of sodium-ion batteries is likely to persistently lag roughly 30 percent behind comparable lithium-ion batteries for that reason alone (Tarascon, 2020). Sodium's potential as a battery material has been known (or suspected) for a very long time. In fact, Jules Verne (1994) imagined sodium batteries as the power source for the Nautilus in his novel *Twenty Thousand Leagues Under the Sea*. Nevertheless, despite a tremendous amount of research spanning at least half a century (Delmas, 2018), sodium-ion battery performance is estimated to be anywhere between ten (Crownhart) and twenty (Freund) years behind the best lithium-ion cells. Of course, some researchers exude a much greater enthusiasm for sodium battery technology than others, and commercial modules are starting to appear (Yahoo! Finance, 2024). The most profitable applications for sodium-ion batteries are likely to be in stationary (electrical grid load balancing) services where their excess weight and volume compared to lithium is less of an impediment than in vehicles or consumer electronics.

Our enthusiasm for more batteries doesn't end with cars. Some electric appliance manufacturers are working to build lithium-ion batteries into all their new models (Coren, 2023a). This push comes as many appliances, particularly induction stoves, draw larger amounts of power than many kitchens are wired to provide. Moreover, a nation-wide switch to all-electric appliances will place additional stress on the grid. Installation of individual appliance batteries might serve to 1) allow appliances to run on lower voltage kitchen circuits, and 2) provide load balancing for the grid by—as with the vehicle example above—selectively discharging the batteries during periods of peak power demand. This all sounds good on paper, but we already face many challenges finding enough new, secure, raw material supplies to meet the projected demands of electric vehicles, consumer electronics, and utility-scale energy storage industries. Adding a battery (or three) to every kitchen will not help solve that problem.

On the other hand, perhaps the temporary unavailability of kitchen appliances performing load balancing services will be less burdensome than vehicles temporarily stranded in their garages. Given a choice between having to wait for the dishwasher to recharge after a hot summer afternoon or being suddenly faced with a dead vehicle battery when I'm late for an appointment, I choose the former. Even so, the fact remains that "demand response," in the sense of borrowing energy from our vehicles, homes, and appliances to replace gaps in power from a grid, is too expensive to deploy everywhere all at once. It also requires a large cultural and behavioral shift. Even if the critical mineral supply problem can be solved—this is not a given, notwithstanding potential new domestic sources (Joselow and Montalbano, 2023)—we will have to learn new habits. That means planning our energy usage to avoid creating bottlenecks that interrupt activities we have long been accustomed to scheduling without giving much thought to them.

Harvey and Gillis (2022) suggest that demand response should be gradually folded into buying, selling, or owning any piece of property by requiring consumers to make incremental energy efficiency improvements. Energy audit reports would, under their plan, become a mandatory part of the seller's disclosure process. Ideally (as envisioned by the planners) homes would not be eligible to be sold without first bringing them up to the latest versions of building codes and energy efficiency standards. In France, an existing house cannot even be put on the rental market without first passing a stringent energy efficiency inspection. Meichtry et al. (2024) reported the story of a Paris apartment owner whose early 1900s era building failed its inspection, forcing him to deduct €50,000 from the eventual selling price. As envisioned for the U.S. market, similar mandatory home repairs might be financed through a second mortgage, or perhaps by loans repaid as surcharges to future utility bills for those who cannot pay cash up front. Imagine being a retiree who has worked his/her whole life to pay off a home, and then being told that the property is now in need of potentially tens of thousands of dollars of upgrades for building code requirements instituted years after the home was built and purchased. This would be a de facto property tax, as well as a direct subsidy for the industries that lobby to ensure the use of their products is enshrined in building codes.

A slightly less invasive example of demand response planning appears in Oregon's new law. It requires all new electric hot water heaters to include a digital interface that enables it to communicate with a remotely controlled demand response system (Oregon Secretary of State, 2023). When activated, the system preheats water at times of low electricity demand, saving both money and grid stress. Water inside the insulated tank remains hot until it is used by homeowners. This sounds like a reasonable accommodation by homeowners, but, as with electric vehicle battery charge sharing with the grid, not all home energy demand is so flexible. Connecting homes more comprehensively to the internet also poses new energy demands for data centers and remote servers. We must also weigh additional requirements for cybersecurity against the energy savings before determining that remote control of home appliances reduces net electrical demand without endangering public safety.

WHAT ABOUT FUSION?

Can the long elusive promise of nuclear fusion sweep away these concerns? Scarcity of metals, transmission wires, and suitable land for development are constraints that would not disappear with the existence of a (hypothetically) limitless source of affordable electrical generating capacity. In particular, the availability of commercial fusion power might make the competition for raw materials even worse, especially in the short run. Nuclear fission, an inverse of fusion that has powered reactors for decades, involves splitting certain large atoms into smaller ones by allowing tiny particles called neutrons to be absorbed into them at low speed. These "thermal" neutrons cause the large atom's nucleus to become unstable, whereupon it breaks apart. In contrast, "fast" neutrons can be absorbed into certain other atoms, transmuting them into different elements (this is a way to manufacture, or "breed" fresh fission fuel). The enormous release of energy results from some of the atoms' mass being directly converted per Einstein's famous $E = mc^2$ formula: the total mass of the split atoms is less than the mass of the larger one. Such mass-energy conversions are entirely compatible with the First Law of Thermodynamics. It turns out, perhaps counterintuitively, that fusing small atoms into larger ones also creates a product having less total mass than the original atoms. Unfortunately, in the case of fusion, the tremendous pressure and temperature required to force small atoms to stick together are so great that, until very recently, no one has found a way to sustain a net positive rate of power generation from the process. Controlling a fusion reaction involves containing temperatures of more than one hundred million degrees Celsius without destroying the reactor vessel (Delbert, 2022). You might say building a fusion reactor is like putting a star in a box, or—more precisely—a giant donut. The gases inside stars are so intensely hot that atoms are stripped of their electrons. The resulting electrically charged vapor is called plasma. A donut-shaped reactor called a tokamak is one of the two methods used to control fusion: powerful magnetic fields generated in the reactor compress and contain superheated plasma in a space small enough to allow the reaction to continue. For this reason, the method is referred to as magnetic confinement. The other leading method to control fusion reactions is called inertial confinement: lasers or extremely high energy x-rays generate shockwaves that compress a tiny pellet of nuclear fuel to the pressure and temperature at which fusion occurs (Rao, 2021).

Magnetic confinement reactors have been built since the 1950s, but no one has ever been able to generate more energy than was required to run them. However, an inertial confinement fusion reactor recently achieved that goal. Energy gain, the ratio of energy production to energy input, of a magnitude greater than 1.0 is the minimum that must be sustained for commercial power generation. The National Ignition Facility in California reported a net energy gain greater than 1.0 in 2022 (Bishop, 2022). That result is indeed a notable achievement for fusion researchers. However, the National Ignition Facility was designed primarily to validate U.S. nuclear weapon stockpile functionality in the absence of underground testing. The economics of commercial power plants will be very different than those in the money-is-no-object world of exotic defense research. Fuel is another constraint. The most feasible fuel for fusion reactors is a mix of two isotopes of hydrogen: deuterium and tritium. Deuterium can be extracted, at significant expense, from water, but tritium normally must be manufactured from lithium. So, add possible future fusion reactors to the list of competing consumers for scarce supplies of lithium alongside electric vehicles, home and utility scale energy storage, and consumer electronics.

The research and development required to make fusion competitive with existing renewable energy options represents a long and heavy lift for the economy. One must factor in the concurrent and extraordinary costs of upgrading the electric grid, keeping up with increased power demand from a growing population, and the price of everything else associated with a clean energy transition. In its favor, fusion would have the advantage of being baseload, or "always on," power. It might face fewer grid interconnection delays than today's wind or solar plants. The amount of radioactive waste generated by fusion reactors would be smaller and shorter-lived than fission waste. But the fact remains that practical fusion power would be neither free nor unlimited in supply. Therefore, fusion is unlikely to be a turnkey, transformative solution to the problems of carbon pollution or universal access to electricity. Popular culture has done us a disservice in this regard in its imaginations of future fusion power. Who can forget the Mr. Fusion device onboard Doc Brown's DeLorean time machine that ran on food scraps and beer in *Back to the Future Part II*? Or the infinitely powerful and totally reliable reactor onboard the Starship Enterprise? Furthermore, while fusion reactions do power the sun and other stars, the electricity we generate from solar panels or concentrated solar farms has absolutely nothing to do with fusion technology. Solar panels rely on the photovoltaic effect, whereby light is absorbed by a material, causing some of its electrons to gain energy and move—resulting in an electric current. Concentrated solar merely focuses heat using mirrors.

Let's be very clear here. Any fusion reactors, if and when they arrive, will just be another way to boil water (or some other heat transfer fluid) and drive a turbine. In this respect they are little different than coal, gas, or conventional nuclear (fission) power plants. Fusion power plants, if built at a scale commensurate with the world's aspirational thirst for power, will almost certainly be designed to consume huge quantities of water. Some envision the electricity generated by fusion to be free and infinite. As we've seen, the water most certainly will not be, and neither will the steel, concrete, copper and other materials needed to build the plants and transmit the power. Desalination could potentially be used to purify brackish groundwater or seawater piped to a fusion plant site. But desalination would represent an additional layer of energy input, too.

IS AN ALL-ELECTRIC SOCIETY A REALISTIC GOAL?

The feasibility of universal electrification rests on our shared expectations of energy availability in the face of many competing demands for capital and natural resources. Stated more plainly: sure, we can use electricity for everything so long as we don't use too much of it. This brings us back to a question of priorities. If artificial intelligence consumes power at the rate of an entire European country, if we are struggling to finance and build new electrical infrastructure, and if the reliability of that infrastructure affects health and safety (as when electric heat pumps are the only source of climate control authorized by building codes), should we allow the market to blindly allocate electricity at a single price for all uses on an equal basis? Should we build power hungry robot friends instead of investing in real human relationships? Should we purchase self-driving cars instead of sharing a bus or train? Do we really need to mine cryptocurrency, or can we just continue to use conventional forms of fiat or commodity money? These questions evoke the memory of controversial discussions over the regulation of other products people may want, but which are often problematic in one way or another: cigarettes; alcohol; marijuana; sugary drinks; social media.

A few facts are worth repeating. Sources of the minerals required to build electronic components are overwhelmingly concentrated in a small number of countries, several of which are politically unstable. Electrical infrastructures in many countries need a substantial overhaul if they hope to continue to supply reliable power for customary applications, let alone support the aspirational goals of "smart grids." Constant advances in computing power will continue to drive an ever-increasing number of electricity-hungry hardware and software applications: a few may be transformative; a great many will be frivolous; the rest will lie somewhere in between. Funding for electrical generation capacity, transmission, distribution, and storage can only come from one of three sources: taxes, public debt, or private usage fees. The first two sources compete for attention with military, social welfare, and other infrastructure priorities; the third relies on wage growth. In the end, the fate of universal electrification appears to require a choice. Do we want continued resource extraction and economic growth with its environmental drawbacks, or will we accept a world where fewer goods and services are produced in the name of delivering those deemed most essential, with the sure to be unpalatable impact on consumers and their lifestyles?

CHAPTER 5

CARBON STORAGE: OUT OF SIGHT, OUT OF MIND?

Do you have one of those closets where you put all the things you don't know what else to do with? Since you can't seem to bring yourself to just quit buying stuff, the closet becomes increasingly overloaded until, one day, you open it to add just a tiny bit more, and everything tumbles out? That's carbon capture, utilization, and storage (CCUS) in a nutshell. Recognizing that society will continue to need combustible fuels, and that the carbon contained in petrochemicals, plastics, and building materials will continue to add to the world's emissions, for the foreseeable future, the hydrocarbon industry is racing to find ways to dispose of carbon other than allowing it to reach the atmosphere. A risk of relying on carbon capture is that much of this carbon could eventually escape, because no entity can be expected to maintain carbon storage sites for hundreds or thousands of years (decades may even be a stretch). The average lifespan of a publicly traded corporation has decreased from fifty to sixty years at the beginning of the twentieth century to less than twenty years today (Gittleson, 2012; Hill et al., 2018; Sheetz, 2017; Garelli, 2023). Political administrations, nation-states, empires, and even, ultimately, geologic structures, will likewise be altered or lost as the millennia pass.

The Intergovernmental Panel on Climate Change (2022) is convinced that limiting global warming to 2°C above preindustrial levels absolutely depends on a combination of future carbon capture and storage (CCS) and existing carbon dioxide removal (CDR). Some of these strategies may involve recycling (utilization) of some carbon in lieu of permanent storage.

> All global modelled pathways that limit warming to 1.5°C (>50percent) with no or limited overshoot, and those that limit warming to 2°C (>67 percent), involve rapid and deep and in most cases immediate GHG [greenhouse gas] emission reductions in all sectors. Modelled mitigation strategies to achieve these reductions include transitioning from fossil fuels without CCS to very low- or zero-carbon energy sources, such as renewables or fossil fuels with CCS, demand side measures and improving efficiency, reducing non-CO2 emissions, and deploying carbon dioxide removal (CDR) methods to counterbalance residual GHG emissions. (IPCC, 2022)

Many carbon storage methods have been proposed. These range from growing millions of acres of trees or other plants to absorb carbon into biological matter, to industrial scale mixing of carbon into the oceans, to injecting billions of metric tons of high-pressure carbon dioxide into underground reservoirs in a process akin to reverse oil or gas production. All the above methods have the potential to be effective, at least to some extent. All are also expensive and technically challenging, and all pose a risk of future carbon escape back into the atmosphere. Ultimately, all will also be necessary and are likely to be deployed together at varying scales in different parts of the world.

Before considering the tradeoffs associated with CCUS strategies, it is useful to recall that nature has its own sophisticated carbon management system. This carbon cycle encompasses activities occurring in the atmosphere, oceans, and geological processes and operates on periods spanning millions of years (NOAA, 2019). The carbon cycle is nature's thermostat. With the increasing concern over emission of greenhouse gases, it is important to realize that the greenhouse effect is an essential enabler of life on Earth. The key challenge now that we have interjected ourselves into the cycle is to maintain the effect at a level that results in a planetary temperature that is not too hot, not too cold, but just right. Without any carbon dioxide in the atmosphere, average temperatures on Earth would fall by around 30°C (54°F) (ESA, 2003), making the planet uninhabitable in many locations for plants and animals. At the other extreme lies Venus, where carbon dioxide is a principal component of the atmosphere and surface temperatures approach 460°C (860°F). Since the amount of carbon dioxide in Earth's atmosphere naturally fluctuates with changes in global climate over geologic timescales, some people question whether the additional increase in carbon emissions humans have engineered as a byproduct of the Industrial Revolution really matters. The answer is yes, and it is a matter of rate as much as of quantity. The natural processes that control carbon emission and storage are imperceptibly slow. At the most simplistic level, we can describe the global carbon cycle as a balance between geological, biological, and chemical reactions. Volcanic activity, animal respiration and ocean outgassing emit carbon into the atmosphere, after which it is slowly reabsorbed into the oceans, consumed by plants, and solidified into minerals. These plants and minerals eventually are buried and transported via tectonic shifts, becoming—millions of years later—a source of fresh volcanic emissions (NASA, 2011).

CAPTURING CARBON

Humans have upset this delicate and slow process by accelerating the emission of carbon to a rate that cannot be balanced by nature's absorbing reactions. Fans of the television show *I Love Lucy* will recall a scene in which Lucille Ball goes to work on an assembly line producing chocolate candy (IMDb, 2023). Things go smoothly until her manager increases the speed of the assembly line, at which point Lucy cannot wrap chocolates as fast as they arrive at her station. Her solution was to eat the candy she was unable to wrap, and it was of course an unsustainable one. At the end she is left looking sheepishly at her manager, chipmunk cheeks stuffed with chocolate. CCUS is no more, and no less, than humanity's grand plan to artificially consume the carbon we are producing faster than nature can convert it into rocks. That is not a reason to abandon CCUS, but it is a reminder that storage will have limits that are not always acknowledged by proponents of the technologies. Like other forms of geoengineering, a topic I will return to in a later chapter, CCUS attempts to allow us to have our cake (candy?) and eat it too.

Capturing carbon before it can be released into the atmosphere is difficult because the concentration of carbon dioxide in industrial exhaust and other waste streams is typically quite low. The job of removing a few percent of carbon dioxide from waste gas is conceptually similar to our discussion of harvesting solar radiation. Recall our earlier metaphor of the red and blue marbles representing usable solar energy and all other incoming radiation, respectively. This time, let black marbles represent carbon dioxide molecules, and green marbles represent everything else in the air—oxygen, nitrogen, argon, etc. At a concentration of approximately four hundred parts per million, you are faced with finding, catching, and separating four hundred black marbles from the 999,600 green marbles that pass through your net every time you hold it up into the air. The problem is a little less daunting if you replace air with the exhaust from a coal burning power plant, where the concentration is closer to one hundred thousand parts per million. Either way, though, it's a slow and expensive undertaking.

Systems for capturing carbon rely on chemical or physical sorting mechanisms that trap carbon dioxide while allowing other gases to pass through. The amount of gas that must flow through the capturing system is many times greater than the carbon dioxide to be trapped. All the gas must also be forced through the system at once by compressors running on electricity, which may or may not come from a low carbon source. Coal and natural gas power plant exhaust contains less than 15 percent carbon dioxide, although chemical plant waste streams may contain as much as 80 percent carbon dioxide (Bettenhausen, 2021). Moving fluids is always an energy intensive and inefficient process. Certain industrial facilities may spend as much as 30 percent of their electricity budgets on compressing air (Stowe, 2017). Only 5 percent to 10 percent of that energy input is actually used in moving the air; the rest is lost as heat. Liquid (mostly water) pumping, which consumes 20 percent of global electricity supply and, similarly, can represent as much as 25 percent to 50 percent of a manufacturing facility's electricity demand, is also woefully inefficient. As much as half of all this pumping energy could be saved by improving equipment design (Hydraulic Institute et al., 2001), but pumps and compressors are always going to impose an unavoidable cost. Carbon capture and storage is fundamentally an exercise in moving fluids, which means that it is certain to be more expensive and inefficient than business as usual. Then add to fluid moving costs the specialized technology to separate carbon dioxide from companion gases. CCUS represents a substantial cost in energy and materials that cannot be eliminated. That makes sense, because it also represents the internalization of carbon pollution (entropy generation) costs. Making polluters pay in this context means explicitly adding carbon capture and storage to the price of all products.

Some of the largest CCUS costs are associated with basic mechanical systems that have been researched and refined for centuries. This is to say that we are not at the beginning stages of the learning curve for them. Carbon capture offers a critique of the notion that learning curves lead to an inevitably downward sloping cost pathway that will make all technologies inexpensive. We will return to this concept of learning curves later when discussing Wright's Law and its implications for the commoditization of energy technologies. The only way to learn and to achieve an economy of scale is to increase production. Increasing carbon capture therefore means increasing carbon production. That in turn means increasing the consumption of fossil fuels, which are doomed to become ever more expensive to locate, extract, and process. Sources of such fuels will move from shallow, free-flowing deposits (like those under the deserts of nineteenth century Arabia) to the ultra-deep ocean, the Arctic, tar sands having the consistency of road asphalt, and rocks that must be shattered using millions of gallons of water (limiting a well's economic lifespan to two to four years). Of course, that is the point of CCUS: It throws a lifeline to the owners of oil and gas assets who might otherwise be forced out of business. Consumers, who would otherwise be forced to curtail their energy usage or change their consumption habits, also get a break.

Again, the lower the percentage of carbon dioxide in an emissions stream, the more difficult and expensive it becomes to trap the carbon. Very low carbon content streams must be processed using specialized chemical catalysts. Carbon rich streams are more amenable to processing using less expensive solvents or physical membranes. Sometimes, however, the easiest method of carbon capture is to simply prevent carbon dioxide from entering the waste stream altogether. In pre-combustion capture systems, hydrocarbon fuels are partially oxidized to produce mixtures of hydrogen and carbon monoxide known as synthesis gas (BGS, 2023). Synthesis gas has been used for over a century as a substitute for natural gas and to artificially manufacture liquid fuels that otherwise would be distilled from crude oil. In regions where oil is not widely available, coal has been used as a feedstock for the production of synthesis gas. In the United Kingdom, mixtures of hydrogen and carbon monoxide volatilized from the abundant Welsh coal fields was piped into homes and used for lighting as "town gas." In early twentieth century Germany, synthesis gas was used, via the Fischer-Tropsch process, to make gasoline and fuel oils for a nation that otherwise would have been entirely dependent on imported oil. In the context of carbon capture, synthesis gas is convenient because the hydrogen in it burns cleanly, producing steam exhaust from which carbon dioxide can readily be separated by condensing the water vapor into liquid.

It is also possible to turn the tables and produce nothing *but* carbon dioxide in the waste stream. Oxyfuel combustion burns the hydrocarbons completely, but in pure oxygen rather than air (which contains roughly 78 percent nitrogen, among other trace compounds). The pure oxygen environment eliminates nitrogen oxides, which are difficult to separate from carbon dioxide. As such, the greater part of the capture expense is in first separating air into oxygen and nitrogen. The separation of oxygen from air is accomplished by chilling air to temperatures below its critical point of -140.7°C (-221.26°F) (Linde, 2019). Refrigerating liquids at such low temperatures is complex, expensive, and energy intensive, but it is a proven technology. Post-combustion capture, in contrast, shifts the cost downstream. Natural air is used as the oxidizer, which means that carbon dioxide must be separated from water vapor, nitrogen oxides, argon, and other impurities after the fuel has been burned. It is in post-combustion capture that the problems of low carbon dioxide concentration are most salient. The only U.S.

power plant to have implemented a commercial post-combustion carbon capture system, the Petra Nova facility in Texas, encountered such high costs and technical difficulties after three years of operation that it was forced to shut down in 2020 (Tracy and Novak, 2023).

Billions of dollars in federal tax credits have been made available to incentivize further attempts at carbon capture. The largest oil and gas companies are hard at work trying to reduce its costs and increase its effectiveness. ExxonMobil alone pledged in 2021 to invest $15 billion in a low carbon business unit that will stand as a peer alongside its core exploration and production operations; a large fraction of this sum will be dedicated to carbon capture research and development (Woods, 2021). This suggests that corporate executives and government policy-makers are betting that significant amounts of oil and gas will continue to fuel society for the many decades needed to pay back those investments. Given the vocal anti-climate change sentiment expressed by the current administration, the odds of this bet paying off are now better than ever.

Physical trapping of carbon dioxide may be accomplished in a variety of ways. Absorption by liquid solvents is perhaps the most technologically mature and economically feasible. Carbon dioxide contamination of natural gas has been a problem in the energy industry for generations. During the earliest days of oil exploration, any gas discovered was merely a nuisance, because without pipelines and a mature customer base, gas cannot be economically stored at large scale. For many years, natural gas was flared, or burned at the well site, its value as an energy source totally unutilized and its carbon emissions ignored. After World War II, the expansion of plastics manufacturing and the desire for energy security led gas-rich nations to build infrastructure to realize the value of gas resources. But there were a couple of carbon dioxide-related problems to solve. First, customers would not pay for carbon dioxide, because it has no value as a fuel (it doesn't burn, because it is fully oxidized). Second, when combined with water, carbon dioxide forms carbonic acid, and this acid corrodes piping and equipment. Thus, the natural gas industry was an early adopter of carbon capture technology in the form of amine absorption. Amines are chemicals that take advantage of the propensity of carbon dioxide to form acids and selectively bind its carbon atoms to nitrogen atoms in the amines. This process frees other components in the gas stream, such as methane, which also contains carbon but does not have the same potential to react with amines, to bubble up and out of the reactor. The carbon-rich amine liquid is then pumped into a vessel where the carbon dioxide is boiled away and piped out for permanent storage or recycling. Exhaust from coal or natural gas power plants, cement or steelmaking operations, or other fuel-consuming factories can likewise be sent through an amine absorption system as currently occurs at many existing carbon capture sites (Hamdy et al., 2021).

Physical membranes, which allow selected substances to pass through while blocking others, are an alternative to liquid absorption systems. Membranes are porous materials, like a sponge or a sieve, that contain small holes that allow some substances to pass through, while blocking others. For instance, you can pour a pot of boiling water and spaghetti together into a strainer, and the water drains out the bottom, allowing you to serve the cooked pasta. Another familiar example of membrane separation is in reverse osmosis desalination equipment: thin polymer films allow water to filter through while salt accumulates on the other side. Similar types of semi-permeable membranes are used to separate carbon dioxide from other gases. The first commercial gas purification membranes were designed to separate hydrogen from mixed gas streams in chemical plants and oil refineries (Han and Ho, 2021). Gas separation membranes can either be designed to isolate molecules above a particular size, or their chemical properties can be optimized to adsorb selected molecules onto the membrane's surface, letting other molecules flow through unimpeded. As with liquid absorption systems or chemical reactors constructed from solid beds of adsorbent particles, much of the energy consumed in membrane-based carbon capture systems is used to pump fluids at the high pressures needed to force industrial volumes of gas through membranes at high flow rates.

All of the above carbon capture systems are technically feasible, if not necessarily cost effective, and all of them have been built. But, by accepting carbon capture—particularly the capture of "new" carbon from combustion exhaust rather than "old" carbon directly from the air

—as a long-term solution to climate change, we incentivize the ongoing consumption of carbon-based fuels and the manufacture of carbon intensive products. Oil and gas are problematic as long-term energy sources not only because they are carbon intensive, but also because they are finite. Carbon capture can play a role in addressing both these concerns if carbon recycling, rather than storage, could be implemented at a scale large enough to supply the global demand for hydrocarbons without additional drilling. Unfortunately, carbon recycling eliminates the principal profit motive for carbon capture among the oil and gas firms that are currently best positioned to invest in it. If a firm cannot accrue tax credits for permanently *storing* captured carbon, then its products made from *recycled* carbon will have to compete with those which do not include recycled carbon, and they will be uncompetitive due to the increased cost of capture/recycling. Consumers have thus far clearly demonstrated that, by and large, they are not willing to pay too much extra for a "green" product. It could ultimately make sense to restructure the oil and gas industry into one that predominantly manufactures synthetic hydrocarbon fuels from recycled carbon and green hydrogen—if compensating policies such as a carbon tax are put in place to allow for the inevitably higher prices of recycled fuels with respect to their conventional counterparts.

Processes that offset emissions using capture and/or storage are termed carbon neutral. If the carbon can be recycled or reclaimed from existing stores, the processes can be made carbon negative. The large-scale manufacture of carbon negative synthetic fuels will require direct air capture (DAC), removal of carbon from the atmosphere itself rather than from industrial facility exhaust. With or without synthetic hydrocarbon fuels, DAC is likely to be a crucial component of the energy transition if we have any chance of ever meeting our decarbonization goals. DAC is even more difficult and expensive than capturing carbon from point emission sources. In spite of this, eighteen small, pilot-scale DAC facilities are currently operating worldwide and eleven more are in advanced stages of development (Budinis, 2022). If all these projects were to be completed, carbon capture rates could reach 5.5 million metric tons per year by 2030, but that would still be less than 10 percent of the rate needed to achieve even carbon neutrality at currently anticipated emissions levels. Implementing DAC at rates of billions of metric tons per year, *plus* widespread adoption of electric vehicles, *plus* massive deployment of solar electricity, *plus* halting deforestation, is considered the only credible way to prevent global temperatures from rising more than 1.5°C by 2050 (Lebling et al., 2022). At this scale, DAC becomes a bona fide geoengineering scheme that must be augmented with increasing amounts of carbon storage capacity—a closet filled to its bursting point with centuries worth of our unwanted waste.

DAC is orders of magnitude more difficult than capturing carbon from power plants. The concentration of carbon dioxide in air is around 0.041 percent, meaning that you are forcing twenty-five hundred times more air than carbon dioxide through the system. These extremely high flow rates suggest designs for DAC systems that pump air through tightly packed tubes of tiny spherical particles engineered to maximize the surface area over which it can flow while simultaneously minimizing pumping pressure. The materials for these particles are highly specialized catalysts with an affinity for carbon dioxide. Chemical reactions occur at the particle surfaces, causing carbon to stick to, or adsorb onto, them. From time to time this chemical reactor must be cleaned and the captured carbon stored or otherwise disposed of, a process called regeneration. Some leading DAC approaches combine solid adsorption with liquid absorption, resulting in a system that is a hybrid between the amine gas systems popular in the oil and gas industry and solid chemical reactors (McQueen et al., 2021).

The global CCUS market is predicted to reach a value of $7 billion by 2030, up from $1.9 billion in 2020. The carbon capture segment is the most lucrative compared to carbon transportation, utilization, or storage (Allied Market Research, 2021). Contrast these figures with Apple's research and development budget, which totaled $4.2 billion in the second quarter of 2019 alone and amounted to 7.9 percent of the company's revenue (Leswing, 2019). Energy companies are currently spending more on developing CCUS technology than they can earn by deploying it, and most of the "revenue" attributed to CCUS projects is in the form of government subsidies. Almost all major electric utilities surveyed in 2023 (Anchondo et al., 2023) were resoundingly pessimistic that they would implement CCUS technology at their power plants anytime soon. The utility industry is particularly sensitive to reliability and cost considerations, as well as customer perceptions regarding those considerations. Even in the face of regulatory pressure, utilities often consider retiring older fossil fueled power plants more economically attractive than installing CCUS.

There is scope for improvement. A geothermal startup in Houston has developed a DAC system that uses geothermal heat to provide just the right amount and type of energy for removal of carbon from air and adsorption onto chemical catalysts (Osaka, 2023b). The temperature of water produced from the company's wells is conveniently optimal for this purpose, making their technology not only economically, but also exergetically, efficient. Recall that exergy is a measure of the maximum amount of useful work obtained by bringing a system from its present state into thermodynamic equilibrium with its surroundings. Geothermal water that is hot enough, but not too hot, to catalyze carbon capture reactions is fortuitously optimal from a thermodynamic perspective. In contrast, if the heat required for carbon capture were to be produced, for example, by using wind turbines to generate electricity and then dissipating it through wires as occurs in a toaster oven, the process would be far less efficient.

GEOLOGIC CARBON STORAGE

Carbon capture coupled with geologic storage (also called sequestration) is touted as pragmatic by many as a comprehensive solution to the problem of excess atmospheric carbon. This combination of processes allows fossil fuel fired power plants to continue their operations and then store their emissions deep underground in geologic formations where they will (ideally) be trapped for thousands of years. This is squarely within the wheelhouse of the oil and gas industry, as it involves drilling wells and injecting fluid into rocks. The process of injecting pressurized carbon dioxide underground is, however, fraught with difficulty. Most of the pore space in the rocks is already occupied by salt water that is pushed outward by the invading carbon dioxide. That water must go somewhere, and its displacement creates extremely high pressures that can cause the surrounding rocks to fracture. This is very similar to the intentional hydraulic fracturing used to release oil and gas from very low porosity shale formations. In the case of carbon sequestration, however, fracturing the storage reservoir is a big problem, because it can cause salt water to percolate upward into fresh drinking water aquifers, making them unsuitable for human consumption. Excessive injection of fluid into underground reservoirs can also lead to earthquakes. This problem, known as induced seismicity, has been well documented over the long history of oilfield and industrial wastewater injection.

For these reasons, the amount of underground rock volume available for safe, practical carbon storage is difficult to estimate. Geophysicists use sophisticated "4-D" seismic measurements that create time-lapsed pictures of the storage volumes below the surface to make estimates. Unfortunately, those estimates are often later discounted by as much as 80 percent to 90 percent (Treviño and Meckel, 2017) after operational data from long-term injection (as opposed to merely testing-phase data) reveal the true limitations of rock fracture strength, proximity to fresh water, and microscopic fluid flow properties. Geophysical measurements provide the size of a box; its length, width, and depth. You also need to know what's in the box. That knowledge only comes to light as you operate wells and take pressure measurements—a process that unfolds over years. Not unsurprisingly, permitting of injection projects is a time-consuming, expensive, multi-year process that must account for and protect against a high degree of technical and operational uncertainty. Only after wells have been drilled, capture plants have been built, and contracts for gas delivery have been signed does the true potential of a carbon sequestration site reveal itself. Carbon capture and sequestration is an important component of the energy transition. However, it will not occur at a speed rapid enough or a scope expansive enough to allow humanity to blithely conduct business as usual or to store all the world's carbon emissions without limit. The cost of geologic sequestration will be similar to the expense of extracting oil and gas, but without the reward of eventual revenue from sale of a product. Tax incentives currently offset some of that cost. But such incentives can't be offered indefinitely lest they become another permanent entitlement program on par with those the agricultural industry enjoys (and, yes, we do pay farmers not to grow crops) (Monnay, 2021).

ExxonMobil announced in July 2023 that it was acquiring a carbon dioxide pipeline company, Denbury Resources, for $5 billion (Eaton, 2023). Denbury had emerged from bankruptcy in 2020, and its valuation three years later is a testament to the enthusiasm oil majors are trying to generate around carbon storage. It may surprise those who are not veterans of the oil industry that drillers often pump as much or more liquid into the ground as they pump out. For one thing, even before the advent of hydraulic fracturing, most of the fluid produced from oil or gas wells was (and still is) water. This "produced water" must be separated from the oil or gas in cylindrical vessels that allow water to sink to the bottom before it drains into tanks. This water, unfit for human or animal consumption, may be treated to a greater or lesser extent, but most of it is reinjected back into the ground. Saltwater disposal wells have been the subject of considerable controversy in recent years, as overuse or misuse of the technology has been demonstrated to cause earthquakes (GWPC and IOGCC, 2021). Induced seismicity effects have likewise been documented in carbon injection wells, which function in a very similar manner. The scientific understanding of induced seismicity has improved dramatically since the number of injection wells began rising after 2015. It is generally possible to construct and operate injection wells safely, but safety costs money and reduces the rate at which fluid can be stored.

The relationships among injection flow rate, subsurface fluid pressure, earthquake potential, and the practical amount of carbon storage are intimately connected. To better understand why, it's useful to expand our prior brief discussion of pore pressure constraints and induced seismicity. Geological reservoirs, as we speak of them, are not lakes of oil and water into which we punch straws and suck like a milkshake—nothing could be farther from the truth. These underground fluid-filled regions are almost entirely solid rock, with microscopic holes (pores) which are saturated with mixtures of oil, water, and/or gas. The pores are never empty, even if the reservoir has been actively exploited to produce oil or gas. Some of the fluid sticks to the surface of the rocks, like that last little bit of peanut butter at the bottom of the jar. The amount of this residual fluid can be substantial. When carbon dioxide is pumped at high pressure into the rock, it pushes the residual water, oil, or gas ahead of it until an impermeable object—like a geologic fault or a lower porosity rock—gets in the way. At that point, continuing to inject fluid will cause the underground pressure to steadily rise, and if injection is not properly controlled or limited, the rocks can break apart. This phenomenon, which is so valuable when trying to produce oil or gas from shale using the technique of hydraulic fracturing, is disastrous in a carbon storage reservoir, because it creates pathways for the carbon to escape. It is, therefore, the proximity of the proposed storage location to natural blockading features such as sealing faults more than anything else that helps to predict how much carbon can safely be injected.

A related issue is the proximity of the storage reservoir to what is known as the crystalline basement, or bedrock. Storage of carbon (or saltwater) very close to basement rock or near faults that connect to the basement has been identified as a predictor of induced seismicity (Chang and Yoon, 2021). The porous geological formations into which we inject, and from which we produce, fluids are sedimentary rocks. These rocks are deposited slowly over millions of years as silt and sand that is gradually buried and compressed until it hardens. Sedimentary rocks lie on top of the tectonic plates of igneous and metamorphic rock that form Earth's continents; these plates act as a "basement" upon which the sediments are stacked. Wherever geomechanical forces push regions of highly compressed rocks together or past each other at angles, massive amounts of energy are stored. Adjacent rock masses become like huge springs waiting to be released, and earthquakes are a manifestation of that release. Even relatively small amounts of fluid leaking between the faces of a fault can reduce the friction and clamping force between them, eventually causing them to slip and allowing the stored energy to suddenly and violently dissipate (Hennings et al., 2021). All other things equal, the deeper a fault slippage occurs, the more and higher frequency energy is released and the more powerful a potential earthquake may become—just as the more tightly you compress a spring, the more forcefully it will rebound once you let go (Abercrombie et al., 2021). Carbon injection programs must be carefully planned and monitored to ensure that displaced water or injected carbon dioxide is not migrating into basement-connected faults. It is an inexact science.

Finally, the rate at which fluid can be injected is a complex function of not just the number of holes, or pores, in the rock, but how the pores are connected to each other—a concept called permeability. Rocks with high porosity but low connectivity between pores are not permeable. This limits the rate at which fluid will flow into and through them and increases the pressure at which fluid must be injected. Increasing injection pressure in turn raises the risk of causing the reservoir to leak carbon, either immediately or far in the future (when fewer people may be paying attention). Engineers always rely heavily on mathematical models derived from theoretical investigations and confirmed by experimental evidence. Laypeople also use models without being aware of it. Every time you step on a bathroom scale and read your weight, you are using Newton's Second Law of Motion: force (weight) equals mass (how big you are) times acceleration (the strength of gravity acting between you and the Earth). You can trust your bathroom scale to work consistently and accurately from day to day. Rock permeability models, on the other hand, are not nearly as reliable in many cases. Moreover, geologists and engineers can't collect and measure samples of all the different types of rocks that comprise a carbon storage reservoir located thousands of feet below the surface. So, the actual behavior of injected fluid is often different from what was anticipated when the reservoir was originally modeled.

The upshot is that the most commonly available carbon storage estimates, based largely on the gross pore volume of rocks indirectly inferred from interpreting the patterns of sound waves, are extremely optimistic. Reported capacity estimates may be orders of magnitude too large in some instances, but investors, government officials, and the public have rarely been aware of that when making plans for how much carbon we can continue to burn while still reducing greenhouse gas emissions. More conservative, dynamic estimates of carbon storage capacity (Núñez-López and Meckel, 2021) can address this shortcoming, to a certain extent, by considering the locations of natural reservoir boundaries and their likely effects on pressure rise. Nevertheless, even dynamic storage capacity estimates are most properly interpreted as wide ranges of values that reflect uncertainty which can only be definitively resolved over time as operational experience provides enough location-specific data.

A broad-ranging study of geologic carbon storage capacity published in 2021 sought to quantify an actionable, economically feasible strategy for storing the ninety-two billion metric tons of carbon dioxide the Intergovernmental Panel on Climate Change considered necessary to maintain global temperature rise below 2°C (Wei et al., 2021). The study identified 2,082 billion metric tons of potential capacity, but it relied upon a static model that will almost certainly prove to be a substantial overestimate of practical storage for the reasons outlined above. As the authors acknowledged, limits on the amount and quality of fundamental geological data that can be obtained in many parts of the world constrain efforts to use more sophisticated models. Bad input data yield bad model results (you may be familiar with the old computer programming adage: garbage in, garbage out). Recalling that the presence of closed geologic boundaries, such as faults, are often the determining factors that limit storage, the data challenge was well articulated by the National Energy Technology Laboratory in their 2013 survey of available storage evaluation models (Goodman et al., 2013):

> In almost all open-boundary method cases the methods do not differ in a statistically significant way in each of the saline formation cases that were studied...Although there are definite differences in the underlying assumptions for open- and closed-boundary methods, for most of the saline formation cases that were studied the variability in underlying parameters was still so great that the estimates of CO_2 storage resource from the closed-boundary method could not be statistically distinguished from those generated by the open-boundary methods. In some cases, however, the open-boundary methods do give statistically significant different results when compared to the closed-boundary method...In general, the uncertainty in the underlying parameters has a much greater impact on

overall estimates of CO2 storage resource than the choice of method does.

In other words, assuming a completely open reservoir volume with no obstructions does not, in most cases, result in statistically significant differences in estimated storage capacity. That is because the rock and fluid properties on which the estimates depend are so poorly understood that adding mathematical sophistication to the model is pointless. Thus, it is difficult to predict to what extent geologic carbon storage can successfully be implemented—even if all other safety, operational, and liability considerations are adequately addressed, and even if a more nuanced, dynamic model of reservoir flow behavior is applied. Assuming, for the sake of argument, that the global storage plan of Wei et al. mentioned above is found to be accurate and feasible, its corresponding cost is estimated to be $8.2 trillion (in 2019 dollars) by 2050. As much as $2.44 trillion of this amount could be recovered if some of the carbon dioxide is used for enhanced oil recovery (EOR). Carbon dioxide injected into an oil reservoir will tend to mix with the oil and reduce its viscosity, allowing more oil to be produced by blending it with carbon dioxide than by using conventional techniques. But the use of EOR naturally implies the longer-term extraction of fossil fuels and thus lowers the net climate change mitigation value of carbon storage.

The enduring success of carbon sequestration will be contingent on finding a way to permanently store or recycle it. The former is conceptually similar to the problem of nuclear waste isolation. Not unlike radioactive material, some of which remains hazardous to human health for hundreds of thousands of years, sequestered carbon must be entombed for times that, from a human perspective, approach the notion of infinity. The sudden release of billions of metric tons of sequestered carbon at some random point in the future would undo our costly efforts to mitigate the effects of climate change. No carbon storage insurance market can be expected to cope with a catastrophic release of that magnitude, any more than fire insurance markets intended to cover periodic, isolated accidents at single homes can cope with a wildfire that abruptly consumes large portions of one of the wealthiest parts (Los Angeles) of the United States. This means that governments are implicitly accepting ultimate liability for the perpetual safety and security of geologic carbon storage.

BIOLOGICAL CARBON STORAGE

There are other ways to store carbon. Organic materials naturally serve as very large carbon sinks, whether plants, trees, soils, or the shells of marine animals. Additionally, carbon dioxide dissolves readily in water—a fact that should come as no surprise to anyone who drinks beer, cola, or seltzer. All the above modes of storage are being actively pursued as additional pathways for large scale carbon sequestration. These natural systems are not technically complex to implement, permitting is typically easier than with geological storage, and they often have ancillary benefits for ecosystems beyond carbon reduction. In contrast to their advantages, these systems have the disadvantage of requiring large amounts of land dedicated—usually to the exclusion of other activities—to long-term storage. Otherwise, the carbon eventually could be released into the atmosphere, creating a sudden liability if offset credits (see below for an example) have been sold against it.

The State of Michigan is an early mover in carbon credit markets. The Big Wild Forest Carbon Project, the first of its kind in the United States, guarantees that the Michigan Department of Natural Resources will actively maintain one hundred nine thousand acres of state forest land as a natural carbon sink, planting and managing the growth of trees subject to third-party verification. In return, DTE Energy will purchase the right to market credits representing the carbon stored in those trees to its natural gas customers (Kart, 2021). DTE will pay Michigan $10 million to exercise this resale right for ten years that began in 2022; proceeds will fund vital state conservation programs. Retail gas customers can check a box on their monthly DTE statement to pay a few dollars per month in exchange for assurances that all carbon dioxide emitted from the gas they burn in their homes has been "offset" by the carbon stored in growing trees. The gas company does not necessarily have to do anything to reduce its own carbon footprint, and neither do its customers: they are deemed carbon neutral. But what happens after the ten-year contract expires? If the trees are cut down for lumber, one might argue that most of the carbon remains trapped in boards and planks built into other structures. The same might be said if dead trees are mulched and plowed into the soil. On the other hand, if a wildfire burns the forest, the offset is negated. No financial settlement can revise the physical history of what happens to the atmosphere—we can only accept the loss and move forward with new carbon storage projects.

A strategy known as Bioenergy with Carbon Capture and Storage (BECCS) melds the cultivation of carbon-rich crops with energy production followed by permanent geologic storage. Trees or food crops and, in some cases, crop waste or weeds can be chemically converted into biofuels or else burned to produce energy (IEA, 2024). In the absence of a permanent storage feature, BECCS can, at best, provide a carbon neutral balance between growing crops or trees and burning biomass. But when combined with permanent storage, there is a potential for actually removing net carbon from the atmosphere (carbon negativity). However, the risk of crop loss, due to drought, flood, disease, or insect plague must also be considered in accounting for the project's net carbon abatement. Land use conflicts are a challenge for all biological carbon storage schemes. BECCS can serve as a compromise approach in terms of land use if "crops" (weeds) harvested for biofuel production or plants used for electricity generation can be grown on land that is less suitable for agricultural production. Corn ethanol production in the United States represents the polar opposite of such an approach. Ethanol has been promoted by the agricultural lobby and heavily subsidized—but it has placed stress on water and food systems and is of dubious benefit in terms of greenhouse gas reduction (Ogg, 2007). Biodiesel produced from palm oil, a European favorite, has led to massive deforestation, loss of habitat, and carbon emissions from burning native trees (Taylor, 2021). On the other hand, Brazil's exploitation of sugarcane for motor fuel production has been slightly more successful and environmentally efficient (Camargo et al., 2022), at least where Amazonian rain forests were not destroyed to make room for the cane fields.

A recent study (Hanssen et al., 2020) estimated that bioelectricity generation from grasses and other non-food crops over a thirty-year period could result in net-negative carbon emissions via BECCS equaling 38 EJ per year, or roughly 32 percent of current electricity demand. However, if only lower quality, abandoned lands are available for cultivation, that number falls to 6 EJ (5 percent). The conflict between BECCS and competing agricultural and urban development needs is evident. Moreover, the full 38 EJ of bioelectricity production would also entail sequestration of 2.5 billion metric tons of carbon dioxide annually—assuming a 90 percent carbon capture rate. The waste or grassland crops studied for use in bioelectricity generation were deemed to be less suitable for biofuel production, with limited potential for negative emissions seen over a thirty-year time horizon. The potential improves over a longer, eighty-year window, but at a lower carbon reduction efficiency than if the crops were used to generate electricity. BECCS will have a role to play in the energy transition, but as with all other technologies, its benefits will come at a cost.

One other way to store carbon with potentially fewer land use conflicts is to do it offshore. Blue carbon, or coastal biological carbon repositories, includes storage in mangrove forests, seagrass meadows, kelp farms, and oyster beds (Macreadie et al., 2021; NOAA, 2023; McVeigh, 2021; Claes et al., 2022). Buried and submerged organic materials have successfully sequestered huge amounts of carbon for eons, just as terrestrial soils do. Of course, there are creatures who live in these areas and farming their habitats will have consequences. Fishermen, beachgoers, and even the military may also object to having long sections of coastline cordoned off for carbon storage. The amount of carbon that can be stored annually using coastal biological systems ranges from 0.4 to 3 billion metric tons, or as much as 7 percent of current annual emissions. Many of these projects could also contribute to the restoration of larger regional aquatic and wetland ecosystems in addition to serving as carbon sinks.

ACCOUNTING FOR CARBON STORAGE RISK

Robust risk profiles that dynamically adjust and acknowledge the great degree of uncertainty inherent in carbon storage are indispensable to building trust in climate management systems. Risk is relevant to all forms of carbon storage, and the types and levels of that risk vary dramatically depending on the nature of the storage medium. Trees are subject to destruction by fire, kelp beds can be poisoned by pollutants dumped near shorelines, and oyster shells can be slowly weakened and dissolved by the rising acidity of seawater. In most cases, the risk we are trying to manage is that of loss, or escape, of captured carbon. Ocean acidification is also an important risk to consider, since its damaging effects occur *while* the carbon remains stored. As the ocean absorbs carbon, it becomes increasingly like club soda in composition. This not only places human made ocean carbon stores at risk, but also works steadily to erode coral reefs and degrade the health of all shell-fish. Carbonic acid is a more insidious corrosive compound than many people realize. The dissolution of carbon dioxide in seawater also increases its density, and this solution will ultimately sink to the bottom of the ocean, where it will react with rocks to form carbonate minerals. A great deal of carbon will—over thousands of years—thus solidify and be buried under the ocean floor.

The risk of carbon *loss* from storage must be quantified for a disparate collection of radically different technologies and media over multiple timescales, knowing that long-term actions by humanity and nature are impossible to predict. We need to have a measure of expected carbon storage—however imperfect—over hundreds, if not thousands, of years if we are to have any confidence in our ability to change the trajectory of the global climate. It is an actuarial problem of epic proportions, and a variety of carbon accounting and reporting plans that attempt to quantify risk have been proposed. Some of these plans take the form of rigorously developed engineering standards that experts periodically update. The Society of Petroleum Engineers (2023) approaches the problem primarily from the perspective of geologic injection and sequestration. Greenhouse Gas Protocol (2023) has been endorsed by the EPA, IRS, and SEC as well as many Fortune 500 companies. The American Carbon Registry (2023) offers standards specifically designed for monitoring nature-based storage. Another document, the PAS 2060 Carbon Neutrality standard (BSI, 2023), attempts to be even more flexible. It not only accounts for operational emissions, but certifies in a transparent, systematic way that the ongoing balance of carbon emitted by renewable products such as e-fuels, BECCS plants, and managed forest carbon sinks is net zero or net negative.

The insurance industry makes these kinds of calculations every time they sell a policy, using data compiled over decades of experience with millions of customers. Given the appropriate kind and amount of data, say on the prevalence of wildfires in a location, a carbon offset contract guaranteed by a Louisiana cypress swamp could potentially command a higher price than the same contracts guaranteed by fire-prone California forests. It is important to standardize, or at least to rationally convert among, approaches used to measure carbon risk because changing them midstream can be fatal for project owners. Wildlife Works, a pioneering biological carbon storage company started by ex-management consultant Mike Korchinsky, is fighting for its life after customers questioned the measures of the efficacy and accuracy of carbon offsets it sold against African nature preserves (Dvorak, 2023b). Even California and the EU, two of the earliest and most enthusiastic advocates of carbon offsetting (via cap-and-trade programs), have called their programs into question. California's Independent Emissions Market Advisory Commission is concerned that carbon removal projects are not actually resulting in reduced emissions (E&E News, 2025). In Europe, the cost of environmental compliance on economic competitiveness is creating huge pressures to rescind or water down emissions reporting requirements (Khan and Mackrael, 2025). In California's case, more rigorous carbon accounting methodologies could help solve the problem; in Europe, more rigorous methodologies could make the problem worse. Environmental stewardship is expensive. Someone has to pay for it.

Any sensible analysis of carbon storage accounts for its true capacity, risk of loss, and cost. The hydrocarbon industry is especially keen to promote capture and storage programs because their business models depend on continued extraction and combustion of fossil fuels. Notwithstanding the Trump Administration's desire to expand American oil and gas production, the industry knows that time is not on its side. The major oil companies are accustomed to making thirty- and forty-year projections of energy supply and demand. Moreover, they realize that the reputational risk from continuing to deny that a global warming problem exists will only grow. Finally, they need to have a credible plan for selling different products in a world where oil is scarcer, or potential carbon taxes reduce demand for it. That does not mean that CCUS is nothing but a cynical ploy, however. Carbon that was emitted hundreds of years ago is still in the atmosphere and, like it or not, we continue to emit more. Removal of at least some portion of this historical pollution must be given its due priority. Unfortunately, to date, most CCUS efforts have failed to meet either economic or technical expectations.

CHAPTER 6
HYDROGEN: A VERY EXPENSIVE NATURAL GAS

Amelia Earhart was the first woman to fly solo across the Atlantic (Michals, 2015). She was not, however, the first professional female pilot. That honor goes to Mme. Sophie Blanchard, a Frenchwoman who flew in hydrogen filled balloons during the Napoleonic Age (King, 2012a). Blanchard's husband spent his life chasing fame and fortune, leaving Sophie with large debts she eventually repaid by succeeding as a public performer and aviation expert. She died in a crash caused when a fireworks display above the Tivoli Gardens in Paris (predictably, in hindsight) ignited her hydrogen-filled balloon. Her tombstone in the Pere Lachaise cemetery features a depiction of her flaming balloon, presumably as a cautionary statement to future aviators. Discovered by Henry Cavendish in 1766, hydrogen was quickly recognized as a potential transportation fuel due to its lighter-than-air properties. The French government even once considered using a fleet of hydrogen-filled balloons to transport troops across the English Channel during a prospective invasion, a plan that Blanchard—who had been appointed chief air minister of ballooning—studied and rejected as impossible.

Global demand for hydrogen in 2021 was ninety-four million metric tons, of which only forty thousand metric tons (0.04 percent) were for so-called "new applications" related to the clean energy transition. The bulk of hydrogen is consumed in oil refineries and for the synthesis of ammonia and other common industrial chemicals (IEA, 2022d). Hydrogen is added to crude oil to remove impurities and to break down the heaviest hydrocarbon molecules into lighter, cleaner, and more useful commercial fuels. Without ammonia-based fertilizers, the world would literally not be able to feed itself at current population levels (UNEP, 2019). The Haber-Bosch process, developed at the beginning of the twentieth century, allowed fertilizers to be directly synthesized from nitrogen in the atmosphere by combining it with hydrogen to make ammonia (UNEP, 2020). Haber-Bosch ammonia production consumes 1 percent of all fossil fuels and generates 1 percent of global carbon emissions (Sanders, 2023). Hydrogen is therefore already an essential component of the energy value chain. But to further scale up hydrogen production to the extent envisioned for global decarbonization could cost tens of trillions of dollars (Tooze, 2023).

HIGH ENERGY BUT DIFFICULT TO HANDLE

The universe is vast, containing as many as a septillion stars. Their contents are almost exclusively hydrogen (NASA, 2023a). The space between stars is not empty—it is also primarily composed of a very dilute concentration of hydrogen gas, approximately one hundred twenty atoms per liter of interstellar volume (NASA, 2020b). One reason hydrogen is so plentiful out there is that there isn't much else with which it can react. Larger elements are produced in fusion reactions that occur inside stars as they age and die. Some of those heavier elements slowly coalesce into planets under the influence of gravitational attraction (NASA, 2023b). Hydrogen is extremely chemically reactive, and on Earth it readily combines with almost anything else. Therefore, it is not found here as a pure element in any significant quantity. It is, of course, widely available as a component of water, the familiar H_2O.

Hydrogen is an excellent energy carrier, but it is more difficult to transport, store, and use than common liquid fuels. It is also non-toxic and, being the smallest atom in the universe, has the highest energy density of any fuel on a mass basis. Hydrogen's extremely low *volu-*

metric energy density, on the other hand, poses one of the greatest barriers to its adoption as a mass-market fuel. To understand this apparent contradiction, recall the four fundamental states of matter: solid, liquid, gas, and plasma. Now imagine a very cold box—cold enough so that whatever is inside the box starts out frozen solid. The volume of the solid is fixed and is independent of the size of the box. If you try to squeeze or push on the solid, it will resist the force of your hand. Now, increase the internal energy (temperature) of the material in the box. It will at some point begin to melt. As a liquid, the volume of the material will be mostly unchanged, but it will no longer be rigid; it will slosh around the bottom of the box or give way when you poke it with a finger. An engineer would say that a liquid differs from a solid in that a liquid cannot resist shear (transverse, or sideways) stress.

Keep heating the liquid. Eventually it will boil, at which point it is in the gaseous state. Gases, like liquids, are fluid—they too cannot resist shear stress. But gases gain a new property that liquids lack: gases can expand to fill the entire space of whatever container they're in. A gas won't stay in a puddle at the bottom of the box but instead will form a cloud that fills the whole thing. This time, the engineer would say that a gas differs from a liquid in that the gas is compressible. An even fancier way of saying this is that the density of a gas is a strong function of the pressure used to contain it. Heat your gas even more, until it approaches the temperature at the center of the Sun. At this point, it has so much internal energy that its individual atoms separate into a vapor of positively charged nuclei and negatively charged electrons that we call plasma. A large amount of the hydrogen fuel in stars exists as plasma, and we can make terrestrial plasma, but hydrogen used as an ordinary fuel is in either the liquid or gaseous state.

This brings us back to the problem of hydrogen's low volumetric energy density. Hydrogen boils at an extremely low temperature (-252.879°C or -423.182°F) under standard atmospheric pressure. If you are at all concerned about the amount you have to pay to run your refrigerator or air conditioner, then you would be appalled by the electric bill sent to someone trying to liquefy hydrogen. Liquid hydrogen takes up less space than gas (has a smaller volume), but the expense of keeping it chilled during the trip from the plant where it is manufactured to the place where it will be used as a fuel often makes the project uneconomic. The alternative is to transport gas, compressing it into heavy, thick-walled tanks. To transport a large amount of energy, the gas pressure must be very high; we must increase its density by a tremendous amount. The hydrogen itself doesn't weigh much (it has low mass), but that is not very helpful because the amount of energy we can use is limited by the size of the tank we use to move it.

Potential synergies between the current market for liquefied natural gas and a future market for liquid hydrogen are worth exploring. An Australian-Japanese joint venture successfully chilled hydrogen gas in Australia, sent a custom-designed tanker ship from Japan to pick it up, and delivered the world's first commercial shipment of liquefied hydrogen to the Port of Kobe in February 2022 (Maritime Executive, 2022). The project has been criticized for using "black" hydrogen (coal as the feedstock for hydrogen production) rather than shipping green hydrogen produced using water electrolysis (Gunia, 2022). The debate over whether to first build a global market for fossil-derived hydrogen or instead channel more research and investment into the production of green hydrogen is a point of tension among proponents of a hydrogen economy. Both tracks are useful at this early stage. Other methods for "liquefying" hydrogen include transforming it into easier-to-transport compounds, such as ammonia or methanol. Hydrogen can also be chemically bound to organic compounds from which it is stripped on delivery to its final destination: these are called liquid organic hydrogen carriers (LOHC). LOHC can react quasi-reversibly with hydrogen, holding it in a convenient liquid state for transport at ordinary temperatures and pressures (Preuster et al., 2017), while also allowing it to be recovered in its pure form thereafter. Suitable LOHC chemicals are often alcohols or hydrocarbons having properties similar to those of gasoline or diesel oil (Niermann et al., 2019). Reversing the hydrogenation reaction is energy-intensive, and the LOHC chemicals must be shipped both ways at considerable cost without directly providing value themselves (like sending a letter in a very heavy and expensive envelope that must be returned to you by the recipient).

There are more than three dozen LNG liquefaction terminals in the planned, approved, or construction phases in the United States (FERC, 2020). Natural gas prices spiked in 2022 due to urgent needs in Europe created by the war in Ukraine (Dezember, 2023). There is nevertheless a risk of overbuilding gas export capacity: European energy networks will eventually stabilize. Even so, EU nations are making big investments in hydrogen in anticipation of long-term fuel switching. A forward-thinking approach for fuel suppliers everywhere would be to make LNG facilities compatible with the future export of liquid hydrogen. Mitsui E&S Shipbuilding, Uyeno Transtech and Yanmar Power Technology have designed a new LH2 tanker class (Marine Insight, 2024) and support for it can be folded into U.S. infrastructure planning. Realization is growing among energy firms about the substantial synergies that are possible from leveraging natural gas liquefaction technology to accelerate the production and transport of green hydrogen (Ramkumar, 2023). A continuing challenge includes selection of the carrier fluid to transport.

Ammonia is a particularly attractive hydrogen carrier, because the infrastructure for making, storing, and using it is sophisticated and available worldwide (Chatterjee et al., 2021). Since it boils at a much higher temperature than hydrogen, less energy is lost in liquefying and transporting ammonia (the loss for pure liquid hydrogen is between 30 percent and 40 percent of its initial fuel energy). However, ammonia is highly toxic and has the potential to pose significant health and safety risks in large amounts (US DHS, 2021). The energy required to recover hydrogen from ammonia is also substantial, although ammonia can in some cases be used directly as a fuel itself. Ammonia can either be blended into the fuel for modified coal-burning power plants, or else used exclusively in purpose-built ammonia-only facilities (Reuters, 2021; Obayashi, 2021).

Moving pure hydrogen gas in pipelines, in addition to incurring the penalty of low volumetric energy density, requires the use of special materials. Hydrogen molecules cause grades of steel commonly used to construct pipelines and associated equipment to become brittle and crack (US DOE, 2023c). The precise causes of hydrogen embrittlement are complex, and experts have considered two overarching theories. The first posits that hydrogen somehow loosens the interatomic bonds within steel, allowing adjoining layers of material to separate and form cracks. The second suggests that hydrogen atoms wedge themselves into imperfections in the crystal structure of steel, creating forces that magnify dislocations within the crystals and allow cracks to propagate. Some researchers support a theory that combines these two physical mechanisms (ANL, 2020). Specialized alloys can, at an additional cost, resist hydrogen embrittlement. Alternatively, pipeline pressures can be reduced to alleviate embrittlement risk. Limited amounts of hydrogen can also be blended into unmodified natural gas delivery systems. As much as 20 percent hydrogen by volume can be blended into natural gas equipment, but the amount is often less depending on the specific end use application (ENTSOG and Hydrogen Europe, 2023). The technologies needed to handle hydrogen safely are known, and the principal barrier to their implementation is, as usual, cost (Frangoul, 2023).

SYNTHETIC FUELS

George Olah, winner of the 1994 Nobel Prize in Chemistry, had a truly brilliant idea for what he called a methanol economy. Methanol is an easy-to-transport liquid that can be used with existing vehicles and fuel distribution infrastructure. Methanol can also be used in fuel cells, it serves as the feedstock for a vast range of plastic and petrochemical products (Olah et al., 2018), and it can function as a LOHC. Methanol has long been used as a niche fuel for high performance cars, especially in racing (Sands, 2015). Its disadvantages include corrosivity to fuel lines and lower overall fuel economy, but it can be used effectively without making radical changes to existing transportation technologies. The trillions of dollars of existing refining, manufacturing, and fueling infrastructure dedicated to supporting our existing chemical fuels cannot simply be thrown into a landfill. Moreover, billions of people worldwide who barely subsist on meager diets and own perhaps a single light bulb and a basic cell phone will not forgo the opportunity to drive used vehicles discarded by wealthy Europeans or Americans. It makes little sense to ignore the possibility of making combustion as clean and renewable as possible, for as long as necessary.

Perhaps Olah's vision would have garnered more attention at the time he presented it if he could have shown it to serve as a powerful nexus of green hydrogen production and CCUS. Recycling captured carbon to make convenient, carbon neutral fuels, plastics, and other products—beginning with synthesis of hydrogen into methanol—is an elegant decarbonization option. It can reduce, or even eliminate, the need to drill for new supplies of hydrocarbon fuel. Just don't expect synthetic gasoline to be cheap. Use of synthetic gasoline, even if derived from completely renewable feedstocks, also does not directly address the other air pollution problems caused by burning hydrocarbons: ground level ozone, nitrogen oxides, and particulate matter. Such synthetic fuels, often referred to as electrofuels or e-fuels, are indistinguishable from their fossil counterparts: they are hydrocarbons in every sense. In the same spirit as Olah's methanol economy, a strategy to synthesize gasoline and diesel oil could "greenify" the well-known Fischer-Tropsch process and its successors and allow existing vehicles, appliances, and transportation networks to continue to operate, but in a carbon neutral fashion. E-fuels could address resource scarcity and energy security concerns associated with imported petroleum to a significant extent as well.

Europe has tentatively embraced e-fuels as part of a plan to ban conventional internal combustion engine vehicles by 2035 (Doll, 2023b). Many EU member countries have for years been adamant that the 2035 deadline was non-negotiable. A projected fifteen-year lifespan of the last gasoline and diesel vehicles sold in that year would ostensibly leave Europe's transport sector carbon-neutral by 2050 in accordance with goals set as part of the Paris Agreement. This was seen as overly aggressive by Germany, which relies heavily on its automotive industry for export revenue. At the time of this writing no final agreement between the European Commission and German negotiators has been reached that would provide an exemption for internal combustion vehicles powered exclusively by e-fuels (Reuters, 2023b). Definitions of e-fuels often refer to biofuels (such as those produced using BECCS), but a potentially more scalable source of e-fuels could be realized through the merger of carbon capture with green hydrogen production. A complementary methanol or e-fuels distribution system existing alongside electrification efforts could relieve strain on our aging power grid and allow us to defer or reallocate funds for some electrical infrastructure upgrades. Dilapidated roads, bridges, water mains, and other public works priorities that compete for financing will not wait patiently for us to complete a clean energy transition; ignoring them will negatively affect productivity and safety. Maximizing the sustainability benefits of synthetic fuels requires us to ensure that their components are indeed renewable. Hydrogen today is predominantly produced from natural gas, in a process called steam methane reforming that is economically efficient and technologically mature. Methane, the principal component of natural gas, is split into carbon monoxide and hydrogen—the synthesis gas we have encountered before—and the carbon monoxide reacts with water to form additional hydrogen and carbon dioxide. In the typical case of "gray" hydrogen, the carbon dioxide byproduct is simply released into the atmosphere. Turning this hydrogen "blue" involves capturing the carbon dioxide and sending it to geologic storage (perhaps if the carbon dioxide is instead recycled the hydrogen could be considered "light blue"). Blue hydrogen represents a legitimate emissions reduction, but it involves the continued production of fossil natural gas and risks the escape of fugitive methane—a potent greenhouse gas in its own right—during transportation. E-fuels—like every other climate change mitigation strategy—are a promising, but not perfect, solution to the problem of atmospheric carbon pollution.

ELECTROLYSIS AND GREEN HYDROGEN PRODUCTION

Splitting water into hydrogen and oxygen, utilizing the oxygen industrially, and delivering the remaining hydrogen as a fuel, can be another way to make combustion carbon neutral or even negative. Electrolysis is the process of electrochemically splitting water. The earliest electrolyzers were similar in appearance to a wet cell car battery (the type you may have, at one time or another, attached jumper cables to).

Solid, electrically conductive electrodes were immersed in a liquid, called an electrolyte, that facilitated the flow of current from one side to the other. Electric current was generated using an external power supply to connect one electrode (called the anode) with the other (called the cathode) across the electrolyte solution and form a complete circuit. A semi-permeable membrane placed in the center of the electrolyte tank allowed ions to move from one side to the other. Water was ionized (split) at the electrodes, with hydrogen gas bubbling up at the cathode and oxygen gas forming at the anode. Modern electrolyzers are more compact, have higher efficiencies, and use a wider variety of electrolytes, but the basic principles of operation remain the same (Yu et al., 2022). Capital costs of electrolyzers range from $500 per kW for the most basic alkaline systems to over $6000 per kW for advanced, solid oxide systems (Corbeau and Merz, 2023).

All commercially available electrolyzer technologies have benefits and disadvantages. Solid oxide systems combine high efficiency with the ability to more smoothly increase or decrease hydrogen production to match the variability of wind and solar electricity generation, but they are less durable than competing technologies. Proton Exchange Membrane (PEM) electrolyzers can also adjust their output to match electricity variability, but they are constructed from expensive materials. Alkaline systems are the most mature electrolyzer technology, but they struggle to adjust to variable renewable electricity output. Anion Exchange Membrane (AEM) technology tries to make up for the shortcomings of alkaline systems, but it suffers from relatively rapid and chronic degradation that reduces the device's operational lifespan (Corbeau and Merz, 2023; Hoffmann, 2002).

Universal challenges of electrolysis used for green hydrogen production include the cost of electricity, the need to ensure that the electricity generated is itself carbon free, and its consumption of large amounts of water. At least 9 kg of water (Beswick et al., 2021) is needed to produce 1 kg of hydrogen (neglecting system inefficiencies, evaporation, and other losses). With annual demand for low-carbon hydrogen projected to reach up to 660 million metric tons (1 metric ton = 1000 kg) by 2050 (Heid et al., 2022), the corresponding water demand would be 5.94 trillion kg per year, equivalent to 5.94 billion m^3 or 2,376,000

Olympic-size swimming pools (Hoefs, 2023). For perspective, this is very nearly equal (100.56 percent of) to the total amount of groundwater pumped (European Commission, 2023a) in the entire drought-stricken nation of Spain (Pleitgen et al., 2023) in 2020.

Over 96 percent of global water is seawater, but commercially available electrolyzers perform poorly with saline water used directly as a feedstock (Farràs et al., 2021). If technology for splitting seawater directly into hydrogen and oxygen without pre-purification (desalination) can be scaled up economically, it could transform the clean energy economy. The cost of desalination is a major barrier to using seawater in green hydrogen production. A 2012 study by the Pacific Institute found that seawater desalination plant operating costs in California ranged from $1.54 to $2.43 per cubic meter (Cooley and Ajami, 2012). This implies a cost between $9.15 billion and $14.4 billion per year to treat enough seawater to supply conventional electrolyzers for use in hydrogen production by 2050. The prospect for reducing this cost is at present unclear. Desalination technology is mature, with the Pacific Institute skeptical that major breakthroughs would be made in the near to medium term.

There are two primary methods available for removing salt from water. The first is to boil it out. Distillation-based desalination is effective, but it requires very large amounts of equipment and energy, because the water must be treated in many steps to arrive at the desired purity. To better understand distillation, recall the classic example of moonshining. Desalination can be accomplished in much the same way, with purified water condensed at the top of a distillation column and salt accumulating at the bottom. Saudi Arabia, the world's largest producer of desalinated water (Reuters, 2020), has used distillation technology for many years, sometimes in concert with a second common desalination method (France 24, 2023), reverse osmosis.

Where distillation operates on principles of thermal equilibrium (the composition of a fluid mixture is a function of its changing temperature), reverse osmosis relies on a combination of mechanical (pressure) and chemical equilibrium. Mechanical fluid equilibrium is simple: fluids flow from regions of high pressure to regions of low pressure. Chemical equilibrium is more subtle. If you take cream with your morning coffee, you have no doubt watched as the creamer spreads in a cloud throughout the coffee. This effect is what engineers call diffusion. Molecules are always in motion, and they constantly collide with their neighbors in a random fashion. Given the opportunity, molecules of liquid in a container will gradually disperse until they are evenly distributed within their surroundings. A slug of creamer poured into the center of a coffee cup will disperse and expand until the concentration of cream is the same throughout the cup (if you don't want to wait you can help the process along by stirring).

In a reverse osmosis desalination system, one side contains fresh water, and the other saltwater. The salt wants to distribute itself so that equal amounts are on each side of the membrane, but it can't because the membrane is impermeable to salt. Establishment of chemical equilibrium then suggests that water from the fresh side of the membrane would move across to help balance the salt concentrations on both sides. But this is not what we want. So, pressure is applied to the salty side of the membrane to force water across in the other direction, which results in the effective dilution of the salt concentration on the fresh side (Smart Water Magazine, 2019). This apparent trick takes advantage of mechanical disequilibrium to reverse the diffusion process. The concentration of salt on the low pressure, fresh side of the membrane is maintained sufficiently dilute to meet the desired water purification standard. Meanwhile, the amount of water on the salty side of the membrane continuously decreases. It can be replenished to allow a steady flow of purified water out of the system.

Electrolyzers designed for saltwater tolerance may someday be able to eliminate desalination as a pretreatment step in green hydrogen production. Sodium chloride (common table salt) in seawater dissociates into chloride ions during electrolysis and causes both toxicity and corrosion concerns (Marin et al., 2023). Many other impurities in seawater, including additional ions, microorganisms, and sediment further complicate using it directly as an electrolyzer feedstock (Gao et al., 2022). An emerging technique to overcome the challenges of using seawater in electrolyzers—particularly in terms of chloride ion contamination—is the application of bipolar membranes. Unlike conventional membranes, bipolar membranes do not facilitate ion transfer across the barrier, but instead direct oppositely charged ions in different directions toward the respective electrodes, where they recombine into hydrogen and oxygen (Mayerhöfer et al., 2020). In fact, bipolar membranes underpin the technique of electrodialysis, a third technology used to desalinate water (Oener et al., 2020). More research is needed to improve the economic efficiency of bipolar membrane-based hydrogen production. Conventional desalination of seawater may ultimately prove to be more feasible, but at a cost that won't be entirely avoidable (Farràs et al., 2021).

Fortunately, there are additional benefits to using desalination as a pretreatment step in hydrogen production because freshwater is a valuable product in its own right. Persistent drought and competition among agricultural, domestic, and industrial users of water are driving more jurisdictions to consider desalination investments. When combined with the potential of hydrogen to serve as a carbon-free energy carrier, systems that can selectively divert water either to human consumption or hydrogen production may prove especially attractive. This is particularly true in coastal communities that have expertise in building and maintaining offshore infrastructure for conventional energy production. Offshore oil and gas infrastructure nearing the end of its economic life may be suitable for repurposing as seawater-supplied hydrogen production facilities, relieving local governments from the responsibility of decommissioning and remediating the sites (Pearson et al., 2019).

Economic feasibility of hydrogen production using electrolysis is highly dependent on the overall design of the system. Important design aspects include: the intended usage of the hydrogen; the manner in which it is supplied as a fuel (Terlouw et al., 2022); its location and proximity to existing transport and distribution infrastructure; and the availability of government incentives. A European-wide study of green hydrogen production opportunities in 2022 estimated that a cost of €3.70 ($4.11) per kg was achievable at that time, with reduction to €2.00 ($2.22) per kg by 2040 likely based on expected efficiency improvements. These figures lie in stark contrast to the U.S. government's "Hydrogen Shot" program goal of producing green hydrogen at a cost of $1.00 per kg by 2031 (US DOE, 2023d).

LOW CARBON GAS TURBINE POWER PLANTS

Hydrogen is also a strong candidate for substitution as a fuel in gas turbine power plants. Proposed federal regulations would require electric power plants to reduce their emissions to levels that would essentially require either the use of carbon capture with conventional fuels, or clean hydrogen combustion (Puko, 2023b). Emission standards under the plan would be indexed to reductions in the costs of those technologies, which means that the standards would gradually become more stringent as industry became more adept at capturing carbon and producing hydrogen. Mixing hydrogen into natural gas power plant fuel, with the goal of gradually replacing natural gas entirely with green hydrogen, could relieve the anxiety of capital markets over financing new plants that might otherwise have to be retired before the end of their useful lives (a.k.a. "stranded assets"). Major gas turbine manufacturers are already selling "dual fuel" units (GE, 2023; Clark, 2023). Some of their customers have begun changing existing natural gas turbine orders to instead purchase flexible versions (Wood, 2021). Disadvantages of the dual fuel approach include the much higher temperatures involved in hydrogen combustion—which could lead to as much as a tripling of nitrogen oxide pollutants compared to natural gas (Kawasaki Heavy Industries, 2023)—and the need to ensure that the entire fuel transportation, storage, and delivery system is redesigned for an eventual switch to 100 percent hydrogen, with the consequent increases in equipment cost.

Despite these limitations, dual fuel hydrogen-natural gas electricity generation seems like an eminently sensible approach. It is nicely consistent with the Inflation Reduction Act's Section 45V tax credit for clean hydrogen and the recently enhanced Section 45Q tax credit for captured carbon oxides. The proposed EPA power plant regulations do not stipulate use of a particular fuel. A major factor that will continue to drive the use of natural gas for electricity production is our need for reliable baseload power. Nuclear power is another carbon free baseload option, but the United States all but completely backed away from permitting new nuclear power plants after 1980 in the wake of the Three Mile Island accident. We will be lucky to commission enough new nuclear capacity to keep up with nuclear plant retirements. Other leading nuclear power nations, including Germany and Japan, followed suit after the Fukushima disaster. On the other hand, China is accelerating its nuclear power program. And, the United States has recently seen a lot of research and development activity focused on novel nuclear power plant designs, the outcome of which will be unclear for some time to come. Montgomery and Graham (2017) predict that—on balance—the world could see a net resurgence in nuclear power by the end of this century. Either way, development of dual-fuel gas power technology makes sense.

In the absence of baseload power, and without large stores of hydrogen fuel, we are basically left with batteries. Batteries can store power on the scale of a few hours, perhaps a few days at the most. Tesla installed what was at the time the world's largest lithium-ion battery to back up a remote power grid in South Australia: it could deliver 150 MW for a duration of 1.2 hours (Wesoff and Gifford, 2020). Dual fuel baseload power plants are being built today that will eventually be able to transition completely to green hydrogen (Puko, 2023a). This fuel can be stored in seasonally relevant quantities in the same manner as our strategic natural gas reserves—in salt caverns or depleted natural gas reservoirs.

HYDROGEN IN TRANSPORTATION

Hydrogen as a transportation fuel has enthusiastic supporters, but it has, to date, failed to achieve critical mass in the marketplace. The hydrogen vehicle technology that comes first to mind for many people is fuel cells, which are in fact another form of electric vehicle. The fuel cell itself is a reverse electrolyzer. In fuel cells, hydrogen (or another compatible compound, such as methanol) and oxygen are electrochemically converted to water (or the equivalent reaction products) and, in the process, generate an electric current. The current generated by a fuel cell is an alternative to storing charge in batteries. Instead of plugging in to recharge a fuel cell vehicle, the driver supplies more hydrogen in a way that is nearly identical to today's gasoline or diesel fueling workflow; connect a pump nozzle and squeeze. The catch, of course, is that hydrogen fueling stations must be available, and very few of them exist today.

The battle for market share between battery electric and hydrogen fuel cell vehicles is reminiscent of the competition between VHS and Betamax, between AM and FM radio, or—for readers of a more recent vintage—between MySpace and Facebook. Large technology platforms rely heavily on network effects to succeed: the more people use a network, the more valuable that network becomes. It is difficult to attract customers to a new platform. But if that platform grows large enough, its dominance is all but guaranteed because the cost and inconvenience of using a less established alternative drives people away from competitors. Owning a hydrogen fuel cell vehicle in 2025 is an act of supreme faith. After a hot summer day of running errands with young children and pulling up to the only hydrogen fuel pump within a hundred miles, imagine that you find it out of order. This means you are facing a tow. And where would you have the tow truck driver deliver your vehicle? There are no other hydrogen stations to help.

In a worst-case scenario, a plug-in electric vehicle owner with a dead battery can at least find a standard household 110 V AC outlet somewhere to recharge—it may take all day, but it will work. You might even carry an extra battery to add a small amount of charge in a pinch. You can't readily carry a supply of high pressure gaseous or liquid hydrogen with the same ease that today's offroad drivers strap gasoline cans to their vehicles for long, remote journeys. Is it possible to build the infrastructure needed for everyone to drive hydrogen fuel cell vehicles? Undoubtedly, yes. But fuel cells today are the Betamax to Tesla's VHS, and a wholesale switch at this juncture seems unlikely. There is a scenario that could possibly change the equation. Heavy duty transportation could adopt hydrogen fuel cells to overcome the weight penalty associated with huge truck batteries, in the process building some of the required hydrogen fueling infrastructure. Small vehicles could access those stations, too. For smaller vehicle owners to make that decision would likely come down to the battery industry's failure to reduce charging time to match liquid refueling, coupled with a perceived improvement in hydrogen fuel availability. Hesitancy among vehicle buyers in the United States to make the switch to electric is proving to be stubbornly durable. High upfront cost, combined with the inconvenience associated with taking anything but short, intracity trips, make plug-in electric vehicle technology a losing proposition for many. Affluent homeowners who drive to work and recharge overnight in the garage remain the most viable EV market segment.

The hydrogen combustion engine is another potential workaround. It is possible to burn hydrogen in a suitably constructed internal combustion engine. Some major vehicle manufacturers, notably Toyota (Qureshi, 2023) and Cummins (Nebergall, 2022) have invested in this technology, which is almost identical to the engines used in conventional cars and trucks. Most research to date (Bekdemir et al., 2022; Krok, 2022) has focused on spark ignition engines—so named because they require a spark plug to ignite the fuel-air mixture before combustion. Diesel (compression ignition) engines have been retrofitted to run on 90 percent hydrogen (Martin, 2022) and the potential for complete replacement of conventional fuels is within reach. Hydrogen fueling capabilities can be added to existing gasoline stations (US DOE, 2009), and refueling time is comparable to gas-powered vehicles—a considerable advantage over current electric vehicle recharging systems. Hydrogen combustion engines would not solve the fuel distribution problems plaguing fuel cell vehicles, but they do offer the potential of breathing new life into engine manufacturing plants, repair facilities, and service stations that otherwise face economic ruin in an age of battery electric vehicles. In addition, retrofits to hundreds of millions of existing gasoline or diesel vehicles might allow at least some fraction of them to run on low-carbon hydrogen. This could defray some clean energy transition costs for developing nations that will struggle to import and maintain a large, new fleet of plug-in electric cars. A hydrogen combustion vehicle strategy would, unlike e-fuels, dispense with carbon capture and recycling and achieve a reduction of the air pollution associated with burning conventional fuels.

Today, hydrogen refueling is expensive and far more difficult for consumers to obtain than conventional liquid refueling. As of 2021 there were fifty-two consumer hydrogen vehicle refueling stations installed in the United States, almost exclusively in California, (Saur et al., 2022)). Hydrogen fuel costs between $5 and $8.50 per gallon of gasoline equivalent (Voelcker, 2022). In contrast, as of 2020 there were approximately 115,000 gasoline stations operating in the United States (24/7 Wall St., 2020)—facilities that often also serve as important sources of basic groceries and household items in their neighborhoods. Electric vehicle recharging that takes place over hours while people wait may not be a feasible use of these businesses, but hydrogen fueling would be. From a consumer workflow perspective, it's a drop-in replacement for gasoline: stop, fill, and go.

Hydrogen refueling can, however, be more difficult for both suppliers and consumers to master than petroleum refueling. Unlike gasoline and diesel oil, which can be delivered at standard atmospheric pressure and temperature, hydrogen must either be transferred to a vehicle as a high-pressure gas, or else at extremely low temperatures as a cryogenic liquid. In both cases, additional risk is incurred by both supplier and consumer. The tanks and nozzles needed for the job are more complex and expensive than those used for petroleum fuels. The quantity of hydrogen fuel delivered to a vehicle—operating with either a hydrogen combustion engine or a hydrogen fuel cell—cannot automatically be stopped using the simple, passive liquid level gauge installed in gasoline nozzles: delivery volume must instead be calibrated using sensors as a function of temperature and pressure. Greater allowances must also be made for the variable heating of hydrogen with changing weather conditions that will increase its pressure inside the vehicle. Sealing of the hydrogen refueling interface against the vehicle is also more critical than with gasoline, due to the much greater volatility of hydrogen and its wider ranges of flammability and explosivity in air. Hydrogen volatility and the potential for leakage also make adequate ventilation of vehicle parking structures, especially home garages, particularly important. That design factor has not necessarily been considered by builders of most existing facilities or homes. All these problems can be resolved, and many experts are confident that hydrogen fuel can be used safely (Hoffmann, 2012). Nevertheless, some tradeoffs and additional costs will be unavoidable.

Transforming electricity to hydrogen via water electrolysis can require as much as three to seven times the total amount of energy as direct electrification of a vehicle (Shrestha and Sun, 2023). Yet, once created, hydrogen is stable and can be stored in seasonal or even multi-year quantities for demand balancing as we do with natural gas and the Strategic Petroleum Reserve. Battery storage is measured in mere hours: one hour of power at 1 MW for a battery capacity of 1 MWh (Bowen et al., 2019). In addition, batteries gradually lose their ability to retain charge. Both Tesla and Hyundai warranty their car batteries to maintain at least 70 percent of their charging capacity (McAleer, 2022), which means you can expect to lose 30 percent of your charging capacity. Battery performance also varies with climate, and where you live can be the difference between your vehicle battery lasting eight years or fifteen years. The prospect of losing a significant fraction of what was already perhaps a marginal amount of vehicle range, or having to spend several thousand dollars on a replacement battery, continues to recommend hydrogen as another viable clean fuel alternative.

Hydrogen represents not merely a fuel or energy carrier, but an entire suite of low carbon technologies having risk as well as promise. Many nations are making large investments in clean hydrogen production, storage, and transportation. There is every reason to expect that the production of green hydrogen will increase over the next several decades. Hydrogen will not, however, replace all other existing fuels or energy delivery systems. At its best, hydrogen will provide the world with badly needed flexibility in its quest to eliminate carbon pollution. The direct use of hydrogen in dual-fuel gas-fired electricity generation seems especially promising in the near term. We turn next to conservation, recycling, and the notion of a circular economy—reusing and restoring resources—and the potential for raw material savings to reduce greenhouse gas emissions.

CHAPTER 7
RECYCLING: NECESSARY, BUT NOT SUFFICIENT

There is no better method of removing carbon from the atmosphere than ensuring that it never gets there in the first place. Conservation and efficiency are excellent and obvious techniques for reducing energy consumption, and reduced energy consumption invariably leads to lower carbon emissions. Manufacturers naturally seek ways to reduce material and energy use, but not because they are trying to reduce resource consumption. Rather, their goal is to improve profit margins and lower consumer costs so that they can sell more products. This is a roundabout way of describing the Jevons paradox: the more efficiently a resource can be produced, the less expensive it becomes, and therefore the more of it people will use overall. Encouraging frugality, on the other hand, doesn't require capital (or government subsidies, or international treaties). It does require people to resist what becomes an understandably automatic tendency to place lower priorities on conserving things that do not cost them in financial terms.

CLEAN ENERGY IS A COMMODITY BUSINESS

Harvey and Gillis (2022) offer Wright's Law, the science behind learning curves, as evidence for their assertion that addressing the costs of climate change will steadily become easier over time. They posit that we eventually will reach a golden age of inexpensive clean technology. There are two issues with counting on learning curves to solve our energy and material supply problems. First, to be precise, following a learning curve to the bottom brings a manufacturer to the lower limit of *fixed cost* as its production volume tends to infinity. In other words, the more of something you make, the more you learn how to make the next one better and cheaper than the last one. And, as the total number of units manufactured rises, the more that the manufacturer's initial investment costs are spread out (or allocated) among them, allowing it to maximize the capital value of its machines. Let's think about this for a moment. If I buy a widget-making machine for a million dollars and it costs me ten thousand dollars to train my employees to use it, the first widget I sell has cost me $1,010,000 to make. The first year my employees only make a hundred widgets, which means I have still spent over a hundred thousand dollars on each one so far. Eventually, they get really, really good at making widgets and can crank out one hundred thousand per year, which I sell for ten bucks apiece. Now I can really earn some money, because I have already paid for my widget-making machine: the cost of each widget now equals the product's *variable* costs— such as packaging, shipping, oil to lubricate the machine, advertising, etc. (I am ignoring here the finer details of amortization, depreciation and other accounting subtleties that would alter this calculation in a real-world context). The fixed costs of manufacturing have been allocated among a very large number of units and have thus become effectively zero. This is basic cost accounting, but in a commodity-driven business it is not an entirely realistic revenue model. Commodity prices are variable costs that can and do rise sharply as the commodity grows scarcer. That is why the cost of crude oil or coffee or bananas never falls reliably and inexorably over time, even as manufacturing efficiency and labor productivity increase. Hurricane floods banana plantations? More expensive bananas. Insects destroy coffee orchards? More expensive coffee. Forced to drill a two-mile long horizontal well thousands of feet below ground using a robot and blast the rocks apart with a water cannon? More expensive oil.

That leads to the second issue: all clean energy technologies are shaping up to become as much commodity driven businesses as oil, gas, and coal are today. Wind turbines, solar panels, batteries: all are dependent on critical material prices, which will rise and become more volatile as supply risk, competition among industries and applications, and awareness of adverse environmental and social impacts from their extraction grows. Rising commodity prices expose a core weakness of the conventional capitalist approach to problem solving. The only way to maintain a low-cost consumer environment is to manufacture huge numbers of product units so that increases in variable costs can be compensated by falling fixed costs. Not only does this exacerbate resource shortages when they occur, but it actively encourages material waste. It is usually better for the factory owner to create millions of copies of a cheap product that he/she can convince people to buy again and again than to make a few copies of an expensive product that lasts for a lifetime. If too many copies of an expensive product languish on store shelves, better to burn them rather than allow them to be sold at a discount (Lieber, 2018). Economies of scale require *scale,* and exclusive brands require *exclusivity*!

Think about the first oil wells. They were drilled where oil was visibly seeping from the ground, such as Oil Creek, Pennsylvania. Early gushers like Spindletop in East Texas could launch pressurized oil geysers over a hundred feet into the air. It was almost too easy, and the resulting cheapness of energy helped lead to displacement of the first electric vehicles by gasoline engines in the early 1900s. Times have changed. Oil and gas, like any extractive commodity, do not follow a strictly downward sloping cost curve: explorers are inevitably driven to more and more challenging locations to recover increasingly lower quality reserves. Today's oil and gas industry is justifiably proud of its ingenuity, but to expect that drilling miles below the seafloor or near the North Pole will ever be cheaper than tapping into the fields of Saudi Arabia at their prime is ludicrous. No learning curve can eliminate the essential truth that depletion inevitably drives commodity markets. Clean energy products that rely on critical minerals (which is all of them) are destined to reach prices that, sooner or later, will be dominated by thermodynamic scarcity.

THE RACE TO FIND MORE LITHIUM

The Inflation Reduction Act has initiated a new gold rush. Intended to stimulate investments in clean energy that would allow the United States to meet its 2050 decarbonization goals, the roughly $370 billion government spending package (White House, 2023b) is a prize. The funds are sought not only by entrepreneurs and corporations, but perhaps even more by small towns that have been left behind by the offshoring of American manufacturing. Colleton County, South Carolina, was understandably ecstatic when they found out in December 2022 that their rural outpost near the Georgia border had been chosen to receive $279 million in investment capital by Kontrolmatik Technologies Energy and Engineering, a Turkish manufacturer of utility-scale batteries (Dvorak, 2023a). The Kontrolmatik plant aims to produce 3 GWh of electric grid storage batteries each year, enough to power 540,000 homes for an hour. Just for a moment, let's pretend that one hour of backup energy storage is sufficient to guard against the worst power outages we can expect during powerful and unpredictable storms. There are slightly more than 124 million households in the United States (U.S. Census Bureau, 2023). You can look at building enough batteries to give everyone in America an hour of energy security in two ways: Kontrolmatik can operate its new factory for the next 230 years, or its investment capital can be increased by a factor of twenty-five (to just under $7 billion) and we can get the job done in a decade. Consider that Kontrolmatik is banking on collecting roughly $1 billion in tax credits as a condition of making its initial investment. This means we have to multiply that figure by twenty-five as well. The bottom line is that a $32 billion public-private partnership would be needed to build enough energy storage to guard against modest power outages. A whole day of energy storage for the nation would run us $768 billion, assuming we could find the raw materials required to build all those batteries. In reality, competition for materials in the wake of such a large battery project would adversely affect the electric vehicle or other industries, causing huge price spikes. This helps explain why China has recently made such a concerted effort to commercialize sodium-ion batteries (Lewis, 2024), a competing technology that—if you recall our earlier discussion—suffers from the chronic and perhaps insurmountable challenge of lower energy density. The energy density penalty, while serious in a vehicle, is less problematic for stationary applications such as electric grid backup facilities. For the time being, however, lithium remains the dominant rechargeable battery metal.

Today's lithium-ion batteries can store between 100 and 265 Wh of energy per kg, with lithium metal comprising between 6 and 11 percent of that mass (US DOE, 2022). Taking the best-case scenario of maximum (265 Wh) energy storage and minimum (6 percent) lithium concentration, a single year of production from the new Kontrolmatik plant in South Carolina would require 679,245 kg of refined lithium metal. If we once again multiply that number by twenty-five to build a nationwide one-hour backup power facility, we obtain just under 17 million kg. To put this number in perspective, as of 2021 there was only a single lithium production operation in the United States, the Silver Peak brine mine halfway between Las Vegas and Reno, Nevada (Dentzer, 2021). In 2019, Silver Peak produced 7.9 million pounds (3.5 million kg) of lithium. The United States would need to increase its 2019 lithium production by a factor of five to support the hypothetical one-hour battery storage program without relying on imports, and we would not be able to use our lithium for anything else—computers, cameras, cars, or cell phones.

The United States already does import a lot of lithium, approximately 2,900 metric tons (2.9 million kg) in 2020 (USGS, 2023b). If we are willing to rely on Argentina, Australia, Chile, and China—the four countries holding 82 percent of the world's known lithium reserves, then we would only need to double or triple domestic production to implement the hypothetical one-hour battery storage plan. But then what about electric cars? The average electric vehicle battery requires about 8 kg of lithium (Orf, 2023). If we replace the existing 1.4 billion cars on Earth with electric versions, we will consume 11.2 billion kg of lithium. This is 659 times the amount required to build America's one-hour battery backup system. The world today has 22 million metric tons (22 billion kg) of economically recoverable lithium reserves, so we should be fine, right? The 335 million strong U.S. population represents 4 percent of the world's eight billion total. If we are all going to live the American Dream, then the world will eventually consume twenty-five times as much lithium as it does now. So in addition to the 11.2 billion kg we'll use to replace our cars (assuming no one else who has never bought a car wants one), we can use another 425 million kg to give everyone in the world a one-hour power outage safety net. If we wanted to increase that to a day of energy backup, we would need an extra 10.6 billion kg of lithium, for a grand total of 21.8 billion kg. We would at that point have consumed virtually 100 percent of the world's current supply of economically recoverable lithium.

It is extremely difficult to replace lithium as a battery material. Even most of the proposed solid-state electrodes, considered the vanguard of battery technology, require lithium (Burrows, 2021). Lithium has the highest charge transfer to weight ratio of any metallic element. In a battery, you need to transfer electrons from one electrode (cathode) to the other (anode). Hydrogen is used in fuel cells because it is highly reactive. It is easy to strip away its one electron and produce ions that are just a single proton, lightweight and easy to move across membranes. Hydrogen isn't the best choice for batteries, but it is certainly a viable candidate material. Before the advent of lithium-ion batteries, nickel-metal hydride (NiMH) batteries were used to power a variety of portable electronic devices, as well as the Toyota Prius (MENAFN, 2022). Other battery designs that, instead of a solid metal hydride, incorporate hydrogen stored as a high-pressure gas have been around for decades and continue to be refined through research (Jiang et al. 2024). The next lightest element is helium, which is extremely stable and won't ionize (so no free electrons are available). Thus, helium is not a candidate for electric energy storage. That brings us to the third lightest atom, lithium. It is a solid metal at room temperature, reactive and electrically conductive, and forms lightweight ions that are easy to transfer electrons from. The charge density of lithium cannot be beat. We have known that about lithium for decades, so, although improved batteries continue to introduce many design optimizations, almost all commercially successful varieties still use lithium. Any improvements in range or charging efficiency that might come from using newer types of lithium batteries are likely to encourage more driving and battery integration into more applications. These could include farm equipment, mining, and long-distance cargo transport (recall the Jevons paradox). The net result may be an even greater demand for lithium than is currently anticipated. Recycling lithium batteries will no doubt increase at some point, but the available processes are toxic, difficult, and will incur an expense that cannot be entirely optimized away (Evergreen, 2022).

AND ALL THE OTHER CRITICAL MINERALS

There is a long list of materials that are crucial to the success of the clean energy transition. The U.S. government maintains a list of such commodities, of which there are currently fifty as of the time of this writing (USGS, 2022). Tracking raw materials was originally a military readiness activity. It has been expanded over the years to encompass industrial input materials that are deemed essential to national competitiveness in civilian sectors of the economy. Some items, like aluminum, graphite, and tin, are not commonly perceived as especially rare. Others, like platinum, are known to be scarce and expensive. Still others, like yttrium, dysprosium, and erbium, are unrecognizable to non-specialists. A few valuable and relatively rare commodities, such as gold and silver, do not appear on the list at all. The list constantly evolves, with shifts in industrial demand resulting in a state of "criticality" being established or eliminated for any material at any time. Copper is a likely future candidate for criticality status in the United States because of its central role in almost every aspect of electricity generation, transmission, and consumption. Copper does, in fact, appear on the list of critical minerals maintained by the International Energy Agency.

We don't have enough copper, and the problem is expected to worsen (Hoyle and Steinberg, 2023). Electrification of transportation, home heating, industry, and all the "smart" technologies that will computerize and control them are expected to drive demand for copper to 36.6 million metric tons by 2031. Meanwhile, supplies are forecast to rise only to 30.1 million metric tons, up from 25.3 million metric tons in 2021 (Khan, 2023a). Mineral exploration and development projects are lengthy and costly affairs. New copper mines routinely take between ten and twenty years to permit and build (Scheyder, 2023). Copper production is currently concentrated in South America, with Chile the largest supplier. In 2023 Goldman Sachs forecast prices to rise 67 percent to $15,000 per metric ton by 2025 (Scheyder, 2023) (a forecast that has, thankfully, been overly pessimistic to date with copper on the London Metals Exchange closing at $9,288 per metric ton as of the time of this writing [LME, 2025]). Copper mining is an invasive operation that can generate air and water pollution, degrade lands and communities, and disrupt animal habitats (Poonia, 2021). Mitigating its worst impacts will naturally increase costs, and those costs will either be passed along to consumers of electricity and other products, or else added to the list of economic externalities that comprise the world's growing environmental debt.

Our raw material problems don't stop there. The world's "supply and investment plans for many critical minerals fall well short of what is needed to support an accelerated deployment of solar panels, wind turbines, and electric vehicles" (IEA, 2022a). One reason is the extreme geographic concentration of their distributions: most known deposits of several critical minerals are located in only one or a few countries. Supply of nearly half of the fifty critical minerals the U.S. government tracks is controlled by the top three producers of each. Not even in the case of oil or gas is this degree of monopolization so high. Critical mineral consumption will need to rise sixfold to approximately 43 million metric tons per year if the entire world is to achieve a net zero carbon footprint by 2050. Even a 100 percent recycling rate for end-of-life products would not entirely eliminate the need for mining new supplies (IEA, 2021b).

The United States today relies on foreign sources for over 90 percent of all critical minerals (University of Michigan, 2023a). Solar energy is often assumed to be cheap, carbon free, and unlimited in supply. However, the technology to capture it is none of those things. Critical minerals needed to manufacture a solar photovoltaic panel include tellurium (75 percent foreign dependence), indium (100 percent foreign dependence), germanium (50 percent foreign dependence), gallium (100 percent foreign dependence), and arsenic (100 percent foreign dependence) (USGS, 2019a). Wind turbine structures rely on aluminum (50 percent foreign dependence). Rare earth elements (90 percent foreign dependence), particularly neodymium, are essential components of the permanent magnets integral to advanced wind turbine electric generators (USGS, 2019a). Batteries are needed for electric vehicles, utility scale energy storage, and consumer electronics. The materials used in rechargeable batteries are sourced from remote and technically challenging deposits: these include manganese (100 percent foreign dependence), lithium (50 percent foreign dependence), graphite (100 percent foreign dependence), and cobalt (61 percent foreign dependence) (USGS, 2019a). One of the Trump Administration's ideas for solving this foreign resource dependence problem has been to unilaterally announce annexation of Greenland (Borenstein, 2025; Kantchev, 2025) and Canada (Morris, 2025) for their minerals—whether they like it or not. These oblique threats made against peaceful neighbors and fellow NATO allies are not only immoral, but also delusional.

MATERIAL COMPLEXITY MAGNIFIES SCARCITY

The mix of elements used in manufactured products adds to mineral consumption. Elemental diversity, the number of distinct elements used in products, has expanded rapidly over the last two centuries, altering the notion of what constitutes a critical mineral in direct proportion to our ability to engineer more complex materials. Rocks, in a largely undifferentiated sense, were the sole critical mineral of the Stone Age, which lasted from roughly 100,000 BCE to 1,500 BCE (Peck, 2019). Prehistoric stoneworkers undoubtedly knew something of the properties of various types of specialized stones, but there was no refining of ores or industrial material production in the modern sense of those terms. Copper and bronze eventually came to dominate tool construction for around a thousand years after the end of the Stone Age. Humans began to separate iron from ore around 500 CE. The Dark Ages slowed the pace of materials development in many parts of the world, but beginning with the Renaissance around 1500 CE, it took engineers only four hundred more years to develop the first modern steel mills. A mere handful of generations later, plastics (1940s), silicon chips (1980s) and biomolecular technology (2010s) expanded the practical definition of critical materials from just a few metals during World War II to literally dozens of elements, spanning nearly the entire periodic table, by 2016.

We are forced to track the supply and distribution of an increasing number of elements to manufacture modern products. They are extracted from an increasingly diverse array of raw materials, located in increasingly dispersed source nations, and embedded in an increasingly complex market of product components. When we embed critical materials in products, a tiny fraction of a percentage of one element can mark the crucial difference between desired performance and failure. This infinitesimal margin makes material accounting such a daunting and underappreciated challenge of the clean energy transition. The case of hydrogen-compatible steel is a (simple) case in point. We cannot be concerned merely with the availability of "steel," but instead must track the availability of alloys that have the properties that resist hydrogen embrittlement. The crystal lattices of simple carbon steels can absorb hydrogen atoms over time, causing the metal to become brittle and crack. More appropriate specialty steel alloy selections for use with hydrogen include AISI 304, 304L, or 316 stainless steels, each of which contains various combinations of iron, chromium, nickel, manganese, molybdenum, carbon, and nitrogen (San Marchi and Somerday, 2014). Therefore, a risk in the supply of any one of the elemental components of these steels can complicate the construction or maintenance of hydrogen-handling equipment. The compositions of steels are well known and globally standardized. Much less certain are the compositions of proprietary products manufactured to privately held specifications by numerous manufacturers who do not necessarily use the same mix of raw materials—even from year to year or batch to batch.

The opaque nature of material specifications further complicates infrastructure planning by making substitution and recycling much greater challenges. Material complexity increases criticality risk even if product functionality improves. Product designers have historically not been trained to explicitly consider material criticality constraints. Weight, cost, aesthetic suitability, durability; these are the perennial concerns of designers. Material supply chain risk, which may emerge years or decades into the product's lifecycle, have not typically been germane to product development plans (Bakker et al., 2019). This is beginning to change because the energy transition is, at its heart, a material transition. Not only is the product design process made more complex by the need to predict future supply chain concerns, but designers are increasingly being asked to balance tradeoffs among longevity, durability, and recyclability: these are impossible to separate from material selection. A product designed for longer service life may contain larger amounts of a rare material, and it may be embedded in more chemically or physically complex forms. This choice may prevent recycling where the critical component(s) are difficult or impossible to separate from a less valuable portion of the product. Likewise, high performance materials that include tiny fractions of a critical element designed to enhance its properties may ultimately be destined for landfills if recyclers have no way to isolate and remove the element at a reasonable cost. Such materials may also cost more in energy terms to manufacture than using lower performance materials to make shorter-lived products (Piwonka, 2002).

THE CIRCULAR ECONOMY

My parents encouraged recycling when I was a kid. We had an aluminum recycling facility near our house, and I collected cans from all over the neighborhood. It was a good source of pocket money for a child in the 1970s and 1980s. I would watch the machines crush huge piles of scrap aluminum into cubes, and it was easy to imagine the metal being melted down and stamped into new cans or tools. We were also encouraged to reuse aluminum foil, extending its economic life in its original form. If we expand the notion of recycling to include reuse in a more general sense, we arrive at the concept of a circular economy. In a perfect circular economy, there is no waste. Every product, every bit of raw material, and every byproduct of consumption finds its way back into use through regeneration or reprocessing. All resources flow in a great, virtuous circle through the economy again and again. It is a nice story, and it is also totally unrealistic.

Actual product circularity involves maximizing the time during which a product is in its state of highest economic value. That is not necessarily always the way to minimize lifecycle energy usage or to eliminate material supply risk. Extending the lifespan of a product may also cause it to not perform as needed under extreme conditions. Repairs can be especially difficult in the middle of a long product lifecycle where replacement parts are not a perfect substitute for the originals. After many years, replacements often can no longer be manufactured, a problem that grows more severe with increasing product reliability requirements. Consider the case of nuclear weapons, a product that is (thankfully) almost never used and, in accordance with international treaties, is not even regularly tested anymore. Most of these devices are decades old, made from materials and parts for which spares have not been commercially available for a generation or longer. To be an effective deterrent against aggression, these weapons must be one hundred percent reliable, twenty-four hours a day, every day, forever. Inspections that discover worn out parts could lead to emergency, one-time fabrication of reverse-engineered mechanisms made from materials that may or may not be an exact match. In this instance, money is essentially no object and the work gets done, but such budgetary freedom will not apply to most products.

So, it remains important in the long run, even in a quasi-circular economy where product lifecycles are stretched to their limits, to mine quantities of virgin minerals wherever possible. Innovation can reduce the amount of raw material needed per unit of energy produced (or vice versa) as methods mature with experience. Technological advances can also in some cases increase the amount of a resource that is considered economically recoverable (a so-called "proven reserve"), but there is inevitably a point of diminishing returns for any resource. As a resource is continuously depleted, it becomes more thermodynamically scarce over a long enough period of extraction from its most concentrated deposits.

Let us return one last time to our marble metaphor. This time, brown marbles are waste rock from a mining operation, and gold marbles are the valuable mineral we're after. The number of brown marbles we have to move out of the way to recover a gold marble starts out small as we begin with the richest deposits, perhaps merely sloshing brown marbles (muddy water) around in a pan to expose a gold marble (nugget). As we consume those deposits, we move on to ore grades of lower and lower quality. This means the number of brown marbles we are forced to discard increases exponentially as we approach the average concentration of a mineral in the Earth's crust (Calvo et al., 2016). Later, when we pivot to massive open pit and underground cave mining, all this blasting and hauling of brown marbles requires more and more energy per gold marble. These mining energy measurements demonstrate most clearly the entropy cost of our insatiable thirst for minerals. For instance, as copper ore grades decrease from 6.5 percent 1 percent, the energy per metric ton required to recover the copper from ore increases from less than 10 GJ/t to more than 100 GJ/t (Calvo et al., 2016). All other things equal, ten times more mining energy means ten times more carbon pollution. Zhang et al. (2021) studied the extraction trends of twenty-five industrially important minerals and found that twenty-three of them experienced increased production between 1920 and 2018, between one hundred and 930 times the initial annual rate. The minerals having the largest production increases were aluminum (930 times), nickel (215 times), and chromium (184 times). Rates of production have accelerated especially rapidly since the late 1990s for most minerals studied.

Turn back once more to the ever-salient example of oil and gas. Conventional petroleum resources, extracted through simple vertical wells from permeable rock formations, have been steadily depleted over more than a century of prospecting, discovery, and production. The shale boom that once again made the United States a primary global producer of hydrocarbons appeared to change that trajectory, but it is a temporary illusion. Today's shales are what geologists call source rocks, because they are the original locations where chemical reactions transformed buried organic matter into hydrocarbons over millions of years. Once formed, the hydrocarbons, being lighter than the saline water surrounding them, gradually percolate upward until they encounter natural traps, such as faults, salt deposits, or impermeable rock layers—a process that requires additional eons to complete. Producing immature oil and gas from its source rocks is the energy equivalent of eating one's seed corn.

Clearly, if we buy less stuff we don't really need or use, we can slow the rate of resource depletion. We can give ourselves more time to redesign products and systems to achieve a greater degree of circularity. It won't solve every problem, but it will help. We may also reduce geopolitical conflict while working to tame carbon emissions. But the temptation among powerful nations is rather to capture and stockpile resources, withholding them from others and eliminating the need for their citizens to sacrifice. That sounds like a great idea if you're on the winning side, but it doesn't matter in terms of climate change whether one country can control critical mineral resources militarily. We cannot isolate our portion of Earth's atmosphere from the rest of the world. We must either collectively accept limits on consumption or the planet's climate will continue to change rapidly in ways harmful to all of us.

NUCLEAR ENERGY AND MATERIAL SCARCITY

It is worth noting at this point that nuclear energy is not immune from any of the material constraints that bedevil other low carbon energy production strategies. Global reserves of uranium—the most common nuclear reactor fuel—stood at 6,147,800 metric tons as of January 2019, an increase of 1 percent over 2017 estimates (NEA and OECD, 2020). Nuclear fuel processing is complex, expensive, and time intensive. Moreover, it is subject to heavy scrutiny due to fears that it will be diverted to weapons use. A 1993 agreement between Russia and the United States called Megatons to Megawatts resulted in the U.S. purchase of 500 metric tons of weapons-grade uranium from Moscow (Hiller et al., 2023). This material was reprocessed into civilian reactor fuel, eliminating the means to build twenty thousand nuclear warheads, but also driving the price of reactor fuel so low that most competing providers were forced out of business. After the original agreement ended in 2013, Russian companies continued to supply the

West with uranium, and the United States still imports up to 25 percent of its reactor fuel from the Russian nuclear energy agency Rosatom.

Global fuel enrichment capacity is slightly greater than current demand, and planned capacity upgrades are expected to satisfy at least some fraction of projected new demand by 2040 (NEA and OECD, 2020). The Biden Administration in 2022 requested $4.3 billion to purchase enriched uranium from domestic producers (Natter, 2022) as an incentive to complete U.S. capacity upgrades. Not all nuclear power plants require large amounts of enriched uranium, however. Canada pioneered use of an alternative reactor design that can be fueled using natural (unenriched) uranium (a design that nevertheless has other disadvantages). So-called "breeder reactors" can use small amounts of enriched uranium to bombard a "blanket" of thorium or depleted uranium with neutrons. Those neutrons are captured by the blanket metal, transmuting some of it into fissionable uranium or plutonium. Moreover, discarded ("spent") fuel rods from existing reactors can be recycled or reprocessed to generate fresh fissionable material. The reuse of spent fuel rods is controversial, because of fears that the process could be diverted to create weapons-grade material for nuclear warheads. The uranium isotope (U-233) bred from thorium is as dangerous as plutonium in that regard. Nevertheless, for many other reasons, including inherent safety of the latest reactor designs and greater availability of thorium *vis a vis* natural uranium, many people advocate a switch to thorium-based breeder reactors rather than the conventional uranium or plutonium fueled reactors we rely on today.

A quick aside: I grew up in the town of Richland, Washington, home of the Hanford Nuclear Reservation. This site was used during the Manhattan Project to manufacture plutonium for two of the world's first atomic bombs (the Trinity test bomb and the Fat Man; the Little Boy used uranium enriched at Oak Ridge, Tennessee [NPS, 2025]). When I was a child no one really knew what his/her parents did for a living. We lived in government-built houses. We called our parents' workplace "The Area." But it was also a place where civilian nuclear power was something to be celebrated rather than feared. The U.S. Department of Energy handed out Simulated Fast Breeder Reactor Fuel Pellets as a way to advertise the energy production potential of our homegrown nuclear technology. Glenn Theodore Seaborg, who won the 1951 Nobel Prize in Chemistry for—among other things—discovering plutonium, gave a lecture one day to my high school chemistry class. My high school classmates voted me Most Likely to Become a Nuclear Scientist. If you really want to see more nuclear power plants built, give every child in the United States an education like that.

Repurposing weapons-grade nuclear material remains a wonderful opportunity to showcase the potential of a circular economy. Biden's State Department disclosed in 2021 that the United States had 3,750 nuclear warheads, down from its all-time high of 31,255 in 1967 (Duster and Gaouette, 2021). Approximately two thousand retired warheads are scheduled for dismantlement. What will become of this highly enriched nuclear material? A revised global treaty to gradually eliminate most of these exceedingly dangerous and unusable weapons could generate fuel to energize a proven source of low-carbon electricity for years. In the absence of such "downblending" of weapons-grade material for use as conventional reactor fuel, unconventional reactor fuel cycles may provide yet other alternate avenues for material supply diversification. In addition to the fuel options noted above, many of the advanced nuclear reactor designs being considered in the United States today specify a fuel that contains only 10 percent to 20 percent enriched uranium, with the balance being a mix of naturally occurring uranium and thorium called High Assay Low Enriched Uranium (HALEU). The use of HALEU in conjunction with small, modular reactors that can be constructed in a quasi-assembly line fashion could reduce supply shortages in the nuclear industry while making the technology safer and more efficient (NASEM, 2023). Permitting and social acceptance will remain formidable barriers to a U.S. nuclear renaissance, however—irrespective of any degree to which the reuse of reactor fuel material and design innovations make the option more attractive.

LAND AS A MATERIAL RESOURCE CONSTRAINT

An old saying among engineers is: if it can't be grown, it has to be mined. All materials of every kind must come from the land. Land (even if it is covered by the ocean) is also where we build energy infrastructure. Infrastructure projects are large, intrusive, and notoriously unwelcome in spite of the benefits they bring to society. Competition for land among farmers, ranchers, homeowners, conservationists, miners, and industrialists may turn out to be the most serious impediment to implementing a clean energy transition—more than mineral shortages, political disagreements, or technological limitations. Land use conflicts are slowing the expansion of wind and solar projects, centerpieces of all credible plans to eliminate greenhouse gas emissions. An interactive map published by the Washington Post (Stevens, 2023c) compares expected land requirements of various electricity generation technologies. Unsurprisingly, wind and solar are the most land-intensive, while a hypothetical all-nuclear grid would occupy the least land. The land-energy bottleneck is a consequence of the fact that, for the first time in history, a proposed energy transition is contemplating a global switch from more dense energy sources to the most diffuse sources of energy.

Proposals for reducing energy-related land use include: 1) more solar, less wind; 2) a preference for building new infrastructure on abandoned mines, hazardous waste sites, and other undesirable plots; and 3) encouraging project designs that can accommodate co-located agricultural uses. These are all potentially good ideas. Another is the preferential expansion of offshore, versus onshore, wind, where high voltage regional power transmission cables can be routed under shallow waters. Subsea power distribution hubs may also be more resilient to storms. Floating wind turbine technology is steadily improving, allowing machines to be deployed out of sight of land. Decommissioning offshore infrastructure (including wind turbines that have reached the end of their useful lives), done properly, can return coastal and ocean areas to their original conditions. The expense of offshore construction will always be greater than its onshore equivalent, but that is a tradeoff we will likely be forced to accept as onshore land constraints grow. In the end, almost all the unprecedented amounts of new infrastructure being planned to wean the world off of fossil fuels is going to be in someone's backyard: it is unavoidable. Population growth, the marginalization of farmland due to climate change, and migration away from flooded coastlines will all drive infrastructure development closer in proximity to human habitation in almost every country.

FROM CONSERVATION TO AUSTERITY

As climate change contributes to increasing amounts of damage and disruption in many locations, the frequency of people with means moving to isolate themselves from those effects at the expense of others will intensify. Climate-driven migration will intensify debates over land use: where to build new homes and infrastructure; where to abandon parcels of land that have become uninsurable due to flood or fire risk; where to resettle climate refugees and how to reform immigration policy. These tensions and the social upheaval they portend have helped justify the economic degrowth movement among proponents. Conservation can be at least a partial antidote to some of these problems, even if it is too late to stop many undesirable changes from taking place for at least some time. To make real, tangible progress toward decarbonization without aggressive acquisition of private land or geopolitical competition for raw materials will demand much, much more than switching to higher efficiency lightbulbs. Conservation in this context will mean real sacrifice of a kind that most of us have never been faced with. To begin to understand this, let us return to the example of transportation using personal automobiles.

A seldom recognized fact that continues to work against our efforts to curb carbon emissions is the counterintuitive American definition of a truck (Van Dam, 2023). We are missing a very large piece of low hanging environmental fruit by not properly incentivizing fuel economy. During the 1960s, the United States imposed a 25 percent tariff (known as the "chicken tax" because it was levied in retaliation for European price controls that hurt American poultry producers) on vehicles classified as "work trucks." Naturally, U.S. domestic automakers took advantage of the new tariff to earn higher profit margins on light duty trucks. As a result, the U.S. market share of sedans has steadily declined. Today, all fifty states are majority "truck." What we now call a crossover SUV will be familiar to those of a certain age as nothing more than the station wagons of our youth, albeit perhaps a bit more stylishly designed. Today's neo-wagons are classified as light duty trucks, and thus, foreign manufacturers of similar vehicles must contend with the anachronistic trade barrier. So today, foreign automakers make cars, and American automakers make trucks. A small sedan, such as the base trim level Toyota Corolla, Honda Civic, or Mazda 3 can get forty miles per gallon on the highway under conservative (law-abiding speed) driving conditions. It is not inconceivable that small cars could improve their performance to a level of fifty miles per gallon or even more. Replacing America's "trucks" with cars would be a significant step toward the mitigation of climate change. The best part of this is that no radical new technology, supply chains, or consumer habits would need to be developed. Additional benefits would include: less wear and tear on roadways and bridges, lower consumption of steel and rubber, and lower costs for consumers. Incentives matter, and we are currently incentivizing wasteful fuel usage.

Americans used to understand what it meant to sacrifice for their futures, for their families, and for the greater good. The intrepid people who first settled what came to be known as the Oregon Territory suffered illness, injury, warfare, and death for the opportunity to own a piece of land and control their own destinies. This does not excuse the treatment of the many indigenous people who were massacred or dispossessed during that period of history. Rather, I tell the story only to demonstrate that desperate people will endure almost anything for the chance to improve their living standards, or those of their children and grandchildren. The threat of climate change has not yet engendered the level of desperation among people needed to make the cultural adaptations that could decisively combat it. Today's energy transition is our generation's Oregon Trail, and there are few wagons lining up to commence the journey. Social change is a real and significant part of any technological transformation. Society must be prepared to accept less flexibility in its consumption habits if we hope to achieve anything close to a sustainable, low-carbon economy. We will have to adjust the schedules of our energy demand, increase the portions of our budgets dedicated to energy, and find creative new ways to conduct familiar activities in ways that reduce our carbon footprints.

Take aviation, one of the most difficult to decarbonize sectors of the economy. Pathways to net zero air travel exist, but they are all expected to increase future costs (Nature Climate Change, 2023). They may also in many cases reduce aircraft range (McKinsey & Co., 2023b). Frequent and same-day transcontinental journeys may turn out, with the benefit of hindsight, to have represented a brief and unsustainable convenience. In centuries past, young European men of means would travel, often with a tutor, for several months across the continent, experiencing a variety of cultures as a capstone experience to round out their educations. Travel was arduous and expensive, even for these children of nobility. Squeezing in a week between semesters at university to have a quick look at Venice or Vienna would have been impossible. These long journeys of discovery were known as the Grand Tour (Royal Museums Greenwich, 2023). The practice is lampooned these days for good reason: the image of rich kids on permanent vacation has what politicians call "bad optics." There is nevertheless an interesting possibility hidden within this anachronistic nugget. Conservation implies doing less in aggregate, but less travel could become less *frequent*, yet more *meaningful* travel.

Reliable global broadband access and the rise of remote work could make egalitarian touring a reality for more people coming from and going to every part of the world. Consider the benefits of giving more workers a sabbatical—an opportunity to travel more extensively, but less frequently—as a means of reducing aviation related carbon emissions and promoting richer relationships and education. Entire families from a broader cross section of society might share the rewards of immersing themselves in new experiences that could blend informal diplomacy, study abroad, and international business. Redefining the pace and style of traveling would also make the realities of climate change, and its effects on our neighbors, more visible and immediate.

This kind of slower travel might be less spontaneous, but it would be more impactful, valuable, and memorable.

Of course, conservation in the context of a sweeping global energy transition must transcend any single economic sector (certainly tourism) and will in most cases not lend itself to pleasant exchanges of one luxury for another. Time, materials, manufacturing capacity, and purchasing power will all have to be diverted to the highest priority activities. The scope of proposed investments in infrastructure threatens to strain government and corporate budgets alike. The magnitude of the task evokes the emergency shipbuilding campaigns undertaken during both world wars. Within days of the United States entering World War I, the government established a National Fleet Corporation. It built the largest shipyard in the world at Hog Island, Pennsylvania (National Museum of American History, 2023). At its peak, Hog Island was building ships at the rate of one every five and a half days from prefabricated parts. Shipbuilding capacity during World War II was even more astonishing. American shipyards launched a mere twenty-three ships during the 1930s, but between 1940 and 1945 a total of forty-six hundred vessels were completed. The San Francisco Bay Area alone built fourteen hundred, or approximately one ship per day over the course of the war (NPS, 2023a). What we did *not* build to accomplish this feat were a great many things, including new cars, home appliances, children's toys, and upholstery nails (US DOD, 2020). The war lasted scarcely four years for Americans. Reducing consumption to a level commensurate with net zero carbon and material sustainability would be akin to mounting a permanent home-front climate war effort in many respects.

There are other things we could do to redirect some of our consumption in helpful ways. Civilian Conservation Corps (CCC) workers were responsible for many of the improvements to parks and other public lands we continue to enjoy nearly one hundred years later. Over a period of nine years from 1933 to 1942, the CCC employed three million single men aged eighteen to twenty-five, many of whom obtained their high school educations and learned trades in the military-style camps where they were also housed and fed (NPS, 2023b). In an era where America's youth increasingly questions the cost and value of a college education, the CCC concept once again holds great appeal. Many young people might enjoy serving in CCC-like programs to learn valuable skills and be directly responsible for helping to mitigate the damages of climate change. At the same time, they could be making the nation's energy systems more resilient, especially in rural or lower income areas. One could view such a program as a domestic Peace Corps. Successful service in the program could also be coupled with a GI Bill-style higher education credit for those inclined to continue their studies. The result would be a cadre of skilled American electricians, cybersecurity experts, solar and wind installers, and related tradespeople making a lasting impact on their neighbors' quality of life and setting an example for the rest of the world to follow.

All resource extraction begins with the easiest to exploit, least thermodynamically scarce sources—the lowest hanging fruit—and becomes progressively more challenging over time. The global recycling rate for lithium is approximately 0.5 percent (IEA, 2021b), not because we fail to recognize the value of lithium, but because the pursuit of economic growth dictates that we use the lowest cost sources of lithium first. That means using freshly mined metal until we have exhausted all viable deposits. Globally, only about 9 percent of all plastic is recycled (OECD, 2023). Mining landfills to recover critical minerals, eliminate toxic wastes from the environment, and free up land for development is a goal that conservationists have chased for years (World Economic Forum, 2017; Lockwood, 2016). Thermodynamic scarcity makes infinite growth of material extraction and consumption impossible. The choices we make about what to build and buy now have consequences that will outlive us.

It is worth having an honest conversation about what is truly important for maintaining our quality of life. Goodwill alone collected 107 million donations comprising 5.7 billion pounds (1.425 billion kg) of used goods in 2021 (Chiu, 2023), much of it clothing. Only half of this clothing is deemed suitable for retail sale, and anything wet or stained is discarded. Less than half of the total weight of usable donations are sold domestically, with the remainder ending up in export channels. In 2021 used clothing markets in Accra, Ghana, were receiving fifteen million used garments every *week* from North America, Europe, and Australia; 40 percent are immediately sent to landfills, which rise like mountains along the country's coastline (Besser, 2021). The same story can be told for cars, home appliances, cell phones, and even ships: a torrent of waste constantly moves from west to east and disappears from affluent public consciousness, leaving fewer raw materials available and contributing to toxic pollution and blight for billions of poorer people. A sustainable economy is, in no small respect, one in which we simply buy less.

Every pound of curbside trash equates to seventy pounds of source production waste (Freinkel, 2011). A great amount of trash is petrochemical-containing material that is incinerated, generating carbon emissions and other pollutants. Reducing initial consumption is more effective than recycling. There is probably no one who makes the case for drastically reducing consumption through a combination of population reduction, lifestyle changes, and efficiency, more convincingly than David MacKay. In his landmark book *Sustainable Energy Without the Hot Air*, MacKay (2009) conducted a bottom-up survey of per capita energy consumption in the United Kingdom. He counted everything from food and retail purchases to transport and industrial output. He then compared this total to the expected capacity of a broad range of renewable energy forms on a per capita basis: wind; solar, tidal; wave; geothermal; biomass. MacKay found no reasonable way to sustain 100 percent of UK consumption entirely with domestic renewable energy. He concluded that at least some amount of economic contraction would be required to balance the energy budget. Every nation's mix of future energy resources will differ, but MacKay's core conclusion that reducing consumption will be essential for sustainability is not likely to change.

THE ULTIMATE LIMITS OF RECYCLING

Europe's enthusiasm for power plants that burn mixed trash, including large amounts of plastic, can be viewed as the lower limit of recycling efficiency; one trip around the reuse circle. Carbon pollution from these plants has not historically been seen as much of a concern compared to the chronically short supply of European landfill space (Gardiner, 2021). Waste to energy is a case of triage in action. Proponents note that, without incineration, land constraints will merely result in the offshoring of European trash to nations where emissions targets are lower. Carbon released into the air contributes the same amount to the global concentration whether it happens in Sweden, Ghana, Canada, Argentina, or the Philippines. Moreover, the dioxins, particulate matter and carbon oxides leaving a waste incinerator are largely invisible and being out of sight, they are also out of mind. A landfill, on the other hand, is large, ugly, smelly, and attracts rats. Politicians do not pay the same costs for allowing the atmosphere to be used as a sewer that they do for occupying valuable land with visible waste.

We can do better, but recycling generates entropy as surely as any other real process. Imagine two bowls, one empty and one full of water with an empty cup sitting next to them. Perform the following experiment as many times as you have patience for: take the cup and transfer the water from one bowl to the other. You will find that there is always a small amount of water in the bottom of the bowl being emptied that you just can't manage to scoop into the cup. This residual water will slowly evaporate between fillings, one molecule at a time, at a rate dependent on how humid your kitchen is. And you will spill a little bit, no matter how careful you are. You might say that you don't need the cup. You may decide to just empty one bowl directly into the other. This is an improvement, but you will still lose water due to evaporation and inevitable spillage. Slowly, but surely, you will lose water in the transfer process at a rate faster than would occur if you just left the bowls alone. This happens because you are generating more surface area from which the water can escape into the air through the action of pouring.

Any chemical or physical process you can invent will suffer parasitic losses of material and energy. The losses will differ in character and degree (and perhaps cost). But they will be there. Humanity cannot expect to steadily grow its population or its per capita standard of living without gradually degrading the availability of its resource base. Concentrated, readily minable ore deposits have a greater degree of exergetic availability than collections of the same metals, plucked one atom at a time from waste streams where they are commingled with myriad other substances. Recycling is useful, but it is subject to diminishing returns when applied over decades or centuries of cyclical recovery. To label recycling as renewability is therefore subjective, and the rate of loss should be considered. Olah's methanol economy might potentially be one of the more efficient industrial scale recycling schemes. Losses from conversion of methanol into carbon dioxide (via combustion) and reconversion of carbon dioxide plus hydrogen into fresh methanol will mostly escape into the atmosphere where they could be subjected to the same capture systems under similar conditions. Recycling is important and efforts to reuse materials of all types should be expanded. At the same time, recycling expectations should be set against achievable targets. Consumption of chemically complex materials will have to be reduced, and we will not be recycling 100 percent of anything.

Land and material use are poised to become ever-tighter bottlenecks in the process of integrating renewable sources into the energy mix. Large wind and solar farms require an order of magnitude more land than comparable natural gas or coal burning power plants (Joselow, 2023). The highest quality, lowest conflict sites have already been spoken for, and wind farm siting seems to be especially controversial. A common complaint is that wind turbines spoil scenic views, whether offshore near beaches or onshore in picturesque rural landscapes. The height of these machines, which increases steadily along with power output, is also an issue. Even the Federal Aviation Administration-mandated safety lights installed on wind turbines to warn pilots are particularly vexing to many people (Najmabadi, 2023). Texas politicians have gone out of their way to block offshore wind farms in state waters of the Gulf of Mexico, ostensibly because they are "an impediment to shipping and fishing" (Baurick, 2024). After all restrictions on development are considered, the amount of installable wind energy in the United States may be as little as 15 percent of its technical potential, covering as little as 10 percent of its potential land area (NREL, 2021a). This problem is not limited to the United States. A large 2023 protest in Norway attempted to block construction of wind farms in areas where indigenous people have traditionally herded reindeer—even the celebrated climate activist Greta Thunberg was involved (Paddison, 2023; AP, 2023). Opposition is also not limited to wind. A proposed $3 billion carbon dioxide pipeline that would carry carbon captured in the Midwest to disposal facilities elsewhere in the country was abandoned after opposition from environmentalists and landowners (Smyth, 2023). Protests over solar farm siting resulted in the cancellation of 1.7 GW of planned U.S. projects in 2021 (Groom, 2022).

Energy production requires land, materials, and capital that must be diverted from other uses. The higher our desired standard of living, the more energy we need and the more materials we will use to produce, transport, and store it. If that energy must also be carbon free, then additional tradeoffs must be accepted. It is unlikely that everyone will be completely satisfied by any feasible solution to the resulting conflicts. Recycling and circular product design are urgently needed, but they cannot entirely replace the need for conservation. The limits to reusability are fundamental. Overcoming the diminishing returns from recovery of increasingly diffuse and embedded minerals will require correspondingly larger energy inputs. This, in turn, will further stress material supply chains. On the other hand, dematerialization of the economy through austerity would require a wholesale reimagining of how we distribute and share resources. People who have worked their entire lives under the rules of our economic system will not take kindly to being told, in their golden years, that they cannot be exempted from rationing, especially if wealthy people are able to use political influence to avoid the pain.

We want very much to believe that, with just a little spit and polish, we can undo the accumulated environmental damage of three hundred years and continue our activities unabated. I really suspect that, on a subconscious level, we are hoping for some kind of collective absolution. There is no doubt that recycling rates for plastics and many critical minerals will increase going forward, particularly if a globally enforceable price of carbon can be established. There is also no doubt that designers will take on the challenge of integrating circularity more tightly into their work by creating products that minimize waste and can be used for as long as possible. There is also, unfortunately, no doubt that these efforts will reach the limits demanded by thermodynamic irreversibility. We turn next to a survey of the markets that underpin energy supply, and the tradeoffs involved in making them work in a carbon free world.

CHAPTER 8
CARBON PRICING AND ENERGY MARKET FAILURES

I n 1503 Pope Leo X was desperately in need of funds to rebuild the Basilica of St. Peter in Rome. An enterprising German friar named Johann Tetzel helped solve the problem by selling indulgences, certificates that served as physical evidence of forgiveness for one's sins—past, present, or even future (Cavanaugh, 2017). Legend has it that a German nobleman once asked Tetzel if he could purchase an indulgence for a sin he was soon to commit. Upon receiving assurances from Tetzel, the nobleman proceeded to beat the priest, exclaiming that his sin was forgiven. Contrition and penance were unnecessary. An indulgence was a purely financial transaction, took immediate effect, and was even marketed as able to release one's dead relatives from purgatory. Tetzel did not invent the concept, but he was a gifted salesman and was recognized by the church for his efforts, eventually becoming subcommissioner of indulgences at Meissen, Saxony. Indulgences featured prominently in the complaints Martin Luther made, which ultimately led to the Protestant Reformation (Justice, 2011). Humanity today finds itself in need of collective absolution for the "sin" of carbon pollution.

Carbon trading markets are designed to serve as emission indulgences. Imagine that I plant a tree, which stores carbon in its leaves and branches. You pay me for this service and then go on to burn natural gas to heat your home. In theory, the two actions are "carbon neutral," their effects on the atmosphere cancel each other. In this way, I serve as your climate confessor, and you are relieved of your personal responsibility for climate change—at least for a while. The workings of carbon trading markets, alongside several other key physical and financial commodity markets, will underpin any future clean energy economy, just as markets have enabled conventional energy distribution for generations. One feature of all such markets is their role in standardization and regulation of the energy or energy-related "product." The sale of bulk energy commodities such as electricity and natural gas requires regulation because their transportation and delivery networks are so massive that it would be prohibitively expensive and wasteful to duplicate them. They are examples of what are termed natural monopolies. Hydrogen, e-fuels, and captured carbon will require similar levels of regulation. Electricity markets will also have to adapt to a world where most power is generated from intermittent sources. Atmospheric carbon is not in and of itself a naturally revenue-producing product. Even so, we will have to account for it with every bit as much care and accuracy as gas or power to guarantee that the avoided emissions do not reappear without consequences, and that the price of managing carbon remains free from arbitrage.

THE IMMEDIACY OF ELECTRICITY GENERATION MARKETS

Asset markets rely on the assumption that any product has a positive unit cost that will increase as people demand more of it. If I own a power plant, I must buy fuel, build generators and wires, and hire people to process payments. The more power my customers use, the more of these things I buy. If there is a shortage of any of the inputs to power production, the price of power will steadily rise. That is the way things are supposed to work. Wind and solar electricity generation disturb this economic balance. They are examples of variable renewable energy (VRE), which is characterized by several unique attributes. First, these power sources do not consume fuel in the conventional sense: solar radiation and wind are practically limitless so long as the sun continues to burn *its* fuel. Second, no one can claim ownership of the sun (yet) or prevent it from shining on any part of the Earth's surface. The same characteristic pertains to wind or any other physical aspect of the environment at large.

The absence of fuel cost as a production constraint leads to a conundrum. If my power plant burns coal, to make more electricity I buy more coal. But if my power plant is solar, and there are fewer cloudy days, I produce more electricity essentially for free. When a solar power plant is designed, engineers estimate how much sunlight will shine around the plant, a task that involves long range, uncertain weather forecasting. Solar plant owners also face a daunting task of estimating how much power people will buy over the thirty, forty or more years of the plant's operation. The fact that the entire economy is moving to electrify almost everything while facing limited supplies of transmission cables and regulatory delays magnifies the uncertainty. As a result, at times a solar plant will produce much less electricity than intended, and at other times it will produce much more. It's the same situation when we estimate demand for wind power.

Overproduction of renewable electricity turns out to be as serious a risk to the stability of power markets as a lack of output. If solar or wind farms are producing more electricity than they can sell or store, and if they lack adequate large-scale energy storage facilities (typically batteries) they face one of two unpalatable options. First, they can curtail their operations—turn off wind turbines or disconnect solar panels, or else ramp down the rate at which generators send their electricity to the grid. Second, they can drop the price at which they sell power, hoping to capture market share from competing plants. Studies of regional power markets across the United States conducted between 2008 and 2017 found that introducing large shares of VRE into a service area led to reductions in average wholesale power prices of as much as $12/MWh (Mills et al., 2019).

Sometimes the less expensive option is to actually drop the price of their power below zero—and pay people to take excess electricity. This relieves the operator of a power plant from spending time and money turning off (and later turning back on) the plant (NREL, 2024). In extreme scenarios of power oversupply, competing plants gradually place lower and lower bids until wholesale power prices temporarily become negative. This price inversion is necessary to keep electricity flowing because the electric grid is nearly as essential to modern life as air or water: it simply cannot fail, ever. Grid operators will initiate rolling blackouts, temporarily shutting down service to small areas one at a time, to relieve voltage or frequency imbalances and protect the larger system from total collapse. Regulators mandate that a certain amount of baseload power capacity—power that, unlike wind or solar, does not fluctuate with the weather or time of day—be continually available at a moment's notice. Maintaining a supply of reserve power requires a greater number of power plants to be operating than otherwise would be needed to meet actual demand. At times of especially high demand, this operating reserve power must be delivered to the grid, preventing the plants from either shutting down entirely or dropping their output below threshold levels (Electricity Advisory Committee, 2019; Götz et al., 2014). This forces other (non-reserve) plants, particularly renewable plants, to reduce their power deliveries instead. This is a cost that is ultimately borne by customers because the average cost of power under such conditions will rise—the reserve plants will have a higher marginal cost of service than the curtailed plants. Conventional coal and nuclear power plants (as well as older, inefficient gas power plants) are more difficult to start up or shut down (a.k.a. ramp-up and ramp-down) quickly than state-of-the-art natural gas, wind, or solar plants, making it very costly for them to curtail their output. Regulators try to respect these physical limitations and will in some cases give slower plants preferential access to the grid. Christophers (2024) describes how J.P. Morgan Ventures Energy Corporation used the slow ramp-up and ramp-down rates of older California gas power plants it controlled to game that state's wholesale power market and make unjustified profits. The company eventually entered into a $410 million settlement with the Federal Energy Regulatory Commission (JPMorgan Chase, 2013). Even under more normal circumstances, the net result of these differences in plant operating characteristics and cost structures is that charging negative prices—paying customers to take their power—is sometimes the only recourse renewable plant operators have.

Coal, nuclear, and other plant operators who purchase fuel tend to lose the most under this scenario initially on a per-unit-power basis, but the longer-term losses for wind and solar operators are more pernicious. If negative pricing occurs often enough, these renewable providers will struggle to repay the initial cost of their plants, because they are often excluded from participating in the reserve power programs under which operators receive a fee in exchange for certifying that a certain amount of "firm" generation capacity will be available twenty-four hours a day on demand (Anderson, 2022). The exclusion arises from the fact that VRE plants cannot generally provide the level of reliability needed to be classified as firm power. Over time, this creates a disincentive to invest in new solar or wind plants, which in turn places a de facto limit on the market share of VRE-generated electricity. At very high fractions of renewable market share (beginning approximately above 80 percent) the cost of adding additional VRE increases nonlinearly because of the grid's inability to adapt to power fluctuations without firm capacity (NREL, 2021b). The challenge is reduced somewhat if nuclear power is counted as "renewable" (nuclear power is carbon free, but it consumes scarce fuel that is no more renewable than crude oil or natural gas).

Utility scale energy storage is one way to avoid negative pricing. Electricity in its pure form cannot be stored—it must be consumed at the very instant it is created. The nature of its consumption can be flexible, however. Batteries consume electricity and store it as chemical energy. Pumps consume electricity and raise water to higher elevations where it is stored as gravitational potential energy in reservoirs. Flywheels consume electricity and store it as the inertia of rapidly spinning objects. Compressors consume electricity and can inject air under high pressures into underground storage caverns. Even more exotic storage schemes have been designed. These include machines that raise large concrete blocks or mountains of trash into the air to store gravitational energy (picture the energy delivered by the proverbial apple falling on Isaac Newton's head, but at a scale large enough to power industrial machinery). Excess VRE that can be stored does not need to be curtailed and can be sold later when power production is lower than demand.

MANAGING ELECTRICITY SUPPLY AND DEMAND

Sustainable power markets must answer the question how to value stored energy. Power that is not stored must be sold at prices that reflect current demand. A certain amount of that power, if stored, might earn the operator more revenue if sold later at a time of higher demand. Taking the concept further, if every power plant were to build sufficient storage capacity to buffer grid fluctuations without curtailment or the need for reserve power, then average prices might rise, adversely affecting consumers to the benefit of generators. This could happen either during nominally off-peak periods (such as overnight) or even at peak demand periods if operators choose to wait for higher prices before selling their electricity. This sets up a game of chicken: prices rise until storage capacity is completely utilized across the network (the time at which price would reach its maximum value). Using storage to hold energy hostage would not benefit consumers, and it would complicate another proposal for stabilizing the grid— demand response.

Demand response approaches the power imbalance problem from the opposite direction. If insufficient generation capacity exists to serve demand, then consumers should be incentivized to reduce consumption before it must be curtailed through outages or sudden reductions (known in the industry as load shedding). The smart meters being installed by retail electric utilities are a first step toward implementing demand response programs. Consumers who fail to voluntarily set their thermostats higher on hot days or lower on cold days can find those settings being automatically adjusted by the utility. More sophisticated implementations of demand response envision giving customers real-time exposure to power prices, rather than charging them a fixed rate. If customers know in advance that running a dishwasher at noon will cost five times as much as running it overnight, more will choose the cheaper option, placing less stress on the grid at the highest demand times. Self-serving uses of energy storage by power providers would skew those price signals, and extreme weather events would transfer the risk of major price spikes to ordinary consumers. Weather-related spikes could result in monthly electric bills of thousands of dollars or force dangerous rationing of life-saving heat or medical device power by people unable to pay. On the other hand, demand response could also alleviate some of the need for storage by encouraging more consumption during periods of excess generation when real-time prices fall. California regulators are moving forward with plans to institute dynamic rates, changes in consumer electricity prices that will incentivize users to shift demand from peak hours to lower demand times (Driscoll, 2023). This will require people to adopt new habits and could result in unequal cost sharing, as people who are unable to shift usage will pay substantially more for power.

CARBON PRICING AND EMISSIONS CREDIT TRADING

Sustainable power markets must also properly incentivize carbon reduction. Carbon emission does not currently have any correlation with consumer prices in many jurisdictions. Carbon pricing of some kind is a necessary adjunct to power markets if we are committed to eliminating coal and other high emitting methods of electricity generation. Pricing carbon pollution can take a variety of forms but can be broadly classified as either taxes or cap-and-trade mechanisms. The merits of a carbon tax versus a cap on emissions continue to be hotly debated. Trading the right to emit carbon by someone who has stored or otherwise avoided an equal quantity of emissions is also controversial. All can be effective, and all can fail.

Carbon taxes are straightforward. Governments that implement a carbon tax charge emitters a fee for each metric ton of carbon dioxide released into the atmosphere. The cost of compliance reduces combustion of fossil fuels and other forms of carbon-emitting activities, just as tobacco and alcohol taxes are designed to discourage consumption of those products. You pay only for the carbon you emit. This seems fair, except that the poor use far greater portions of their incomes to meet basic energy needs. For that reason, carbon taxes are seen by some as regressive. One potential solution is to exempt certain levels or types of carbon consumption: for example, taxing only gasoline purchases above a threshold amount. Another solution is to redistribute revenue from carbon taxes as a rebate to low-income households. This would recycle the money through the economy and transfer funds from wealthier to poorer consumers. A downside is that both fixes encourage greater carbon consumption by a larger fraction of the population. Lobbyists also will relentlessly carve out exceptions for their clients' products and dilute the tax's climate effectiveness.

Cap-and-trade schemes are somewhat more complicated. A government implementing cap-and-trade sets an initial limit on the total amount of carbon that can be emitted in its jurisdiction. The limit is allocated to industrial facilities as a set of carbon credits that are "spent" by conducting business as usual. The credits are sometimes issued to firms at no cost. But firms may instead be required to purchase their carbon allotment at auction—especially where they fail to meet operational emission benchmarks (European Commission). If a business can partially decarbonize its operations, it may sell surplus carbon credits to competitors or other entities, who in turn are authorized to emit more. Overall emissions are gradually reduced by lowering the number of credits issued by the government each year until a final target has been reached. Ideally, the steady ratcheting down of allowable emissions reduces the influence of lobbyists, but the system is still vulnerable to failures in a couple of ways. First, the program can be compromised by exempting from participation certain sectors of the economy, say cement manufacturing or steelmaking, on the grounds that compliance is "too hard." Second, targets that are overly generous allow a business to bank credits to an extent that can lower the price of future allowances enough to eliminate decarbonization incentives.

Some policymakers nevertheless encourage purchasing surplus emissions credits as an effective form of carbon reduction. Carbon offset markets attempt to make polluters pay by encouraging firms that can avoid emissions to sell the right to emit an equal amount to another firm. In theory, this leads to carbon neutral operations on an economy-wide basis. Offset markets can be an effective policy tool, but they can also be used to transfer pollution to poorer countries having lax environmental rules. The price of credits also must be lower than any fine a firm would have to pay for just selling more of its carbon-intensive product at market prices, yet high enough that a low carbon producer can compensate for the costs of its more expensive product. Moreover, it can be extremely difficult to account properly for the quantity of offsets that have been generated, or to adjust those amounts for carbon losses that may occur over time due to sequestration leaks, forest fires, or other accidents.

California's Low Carbon Fuel Standard (LCFS) program is one of the world's most mature carbon offset markets. Even so, in 2023 it began to experience instability (Henderson, 2023). This was not unexpected. The state stipulates that a minimum fraction of all fuels must meet renewable energy criteria. Demand for credits will remain strong until low carbon fuel supply reaches the required level. After that, the incentive for conventional fuel producers to purchase credits will fall. Another way to stimulate demand for credits is to tighten the fuel standard measured as an average carbon intensity for all fuels sold in the state. Recent moves by the California Legislature to lower this carbon intensity target have prompted Chevron to take pains to explain to angry customers there that rising gasoline prices are directly related to LCFS mandates—or the inability to meet them (Rosas, 2025). LCFS credits are a government payment designed to be equal to the social cost of carbon, or the cost of pollution-related environmental, health, and property damage attributable to raising the atmospheric carbon concentration by a certain amount (typically one metric ton). Increasing the minimum fraction of renewable fuel required (or reducing the average carbon intensity target) can temporarily boost credit demand. The limit will obviously be a 100 percent renewable fuel mandate, because if all fuel sold is renewable then no one will need credits.

COUNTING CARBON ATOMS

Robust accounting systems that can accurately track the masses of carbon emitted, stored, and traded across time and space will make the difference between effective markets and greenwashing. The airline industry is eager to promote carbon offset trading, because tourism is responsible for 11 percent of all carbon emissions, and because air travel contributes the largest share of that amount (National Geographic, 2023a). More than a third of American travelers surveyed in 2018 were willing to pay higher ticket prices in exchange for a guarantee from the airline that carbon emissions associated with their trips would be offset. There is evidence that airlines are failing to sequester quantities of carbon equal to the offsets they are selling to customers (Shadel, 2023). The International Air Transport Association claims that all offsets are evaluated using criteria mandated by the International Civil Aviation Organization (IATA, 2025). But these criteria pertain only to the aviation industry, and other industries have different criteria which may or may not be equivalent. From an overall economic perspective, carbon accounting systems are fragmented, lack a common basis for assigning offset quality (Huber et al., 2024), and therefore are not as reliable as they need to be.

All greenhouse gas reporting is further complicated by the need to separately account for emissions stemming from three distinct source categories: an entity's direct operations; the producer of the energy purchased by that entity; and the entity's upstream and downstream value chain (vendors, customers, etc.). These three emissions sources are referred to as Scopes 1, 2, and 3, respectively, classifications used by virtually all public and private emissions management systems to track carbon (US EPA, 2023b; US SEC, 2022; Council on Environmental Quality, 2016; Greenhouse Gas Protocol, 2023; Corporate Finance Institute, 2023). The scope of emissions is relative: everyone's Scope 2 or 3 emissions are someone else's Scope 1 emissions. Because Scopes 2 and 3 tend to be the largest sources of emissions, and because those scopes are not under the direct control of a reporting entity, mandatory greenhouse gas reporting is a controversial idea (Mufson, 2023). Corporations naturally worry about the cost of compliance, but comprehensive reporting of emissions is essential to establish fully functional carbon trading markets. Emission reports must also include storage of carbon and data on year-over-year changes. As more products fall under greenhouse gas reporting mandates, assessing emissions contributions along their supply chains will become simpler and more accurate. Eventually, in the limit of economywide reporting compliance, the relative scoping challenges would largely disappear.

Adjusting for the risk of carbon loss from diverse storage schemes (such as trees, geologic sequestration, and kelp farms) is particularly important to maintaining reporting accuracy. Quantifiable, standardized risk metrics will enhance emissions reports as tools for both financial valuation and carbon offset trading. Risk adjusted carbon accounting is conceptually similar to mineral reserves tracking. The economically recoverable quantity of a mineral is typically far less than the physically available quantity. Geologists may have strong evidence that a vein of gold, silver, or copper exists in a particular location, but mining engineers may not have the technology to extract it at a reasonable cost. By the same token, the expected value of stored carbon must be discounted in the face of risk. This is the inverse problem from mineral mining: the risk associated with storage is that the technology used to keep carbon *from being extracted* may be inadequate or exorbitantly expensive. Examples of carbon loss risk include: a wildfire releasing carbon from a managed forest; seismic activity that compromises geologic reservoir integrity; and land development projects that eliminate plant-based carbon sinks. Each of these risks differs in probability, magnitude, and the potential for reversal or remediation. One cannot unburn a wildfire, but if carbon sensors at the ground surface indicate that a leak may have occurred in a geologic storage repository, engineers can attempt to drill wells nearby to produce water and draw carbon dioxide away from the source of the leak. Without a clear understanding of the probability of carbon escaping back into the atmosphere and the ability (if any) to arrest that process, it is difficult to place consistent values on emissions credits.

The challenge of carbon valuation also raises questions of liability. To effectively counteract climate change, carbon storage must be, for all intents and purposes, permanent. Permanence is an unwieldy concept from the standpoint of writing contracts. Very few human organizations have endured for even as long as a few centuries: empires rise and fall; companies are founded and dissolved; religious trends wax and wane. Who will stand ready to take responsibility for the escape of carbon stored one hundred, five hundred, or one thousand years earlier? What form would such restitution even take? Carbon credits are designed to allow someone else to emit greenhouse gases that the seller chose *not* to emit. The only climate-relevant remedy for lost carbon storage is new carbon storage. The fallacy of issuing monetary insurance policies as compensation for global environmental degradation that will have consequences for centuries has led some environmental advocates to question the wisdom of allowing carbon offsets at all (Al Ghussain, 2020).

A neutral, transnational organization, such as the World Bank, might be chosen to hold a global escrow account to cover at least some types of carbon loss consequences or defaults on obligations to offset emissions. Payments countries make to the account could be calculated based on the nation's historical and projected future emissions and the projected costs of adapting to a changing climate (a moving target, to be sure). Such costs might include relocations of people displaced by fires, floods, crop failures, or political instability; seawalls, coastal pumping stations, and other sea level rise countermeasures; and subsidies or food vouchers for nations no longer able to grow crops in previously productive agricultural regions. If escrow payments are high enough to incentivize more robust carbon storage methods, then some losses may even be prevented before they occur. Establishing a truly global, transparent, risk-adjusted price for carbon is the only way to equitably internalize the costs of pollution across national boundaries.

Regional efforts along these lines have met with modest success, particularly in the European Union. The EU Emissions Trading System (ETS) was established in 2005 and is now in its fourth phase of strengthening (European Commission). The number of available credits will decline by 2.2 percent per year until 2030 (EC, 2020), up from 1.74 percent per year during Phase 3. Firms the EU required to participate in ETS decreased their emissions by 35 percent in aggregate between 2005 and 2021. For its part, California's LCFS has been recognized for reducing the carbon intensity of the state's transportation sector by 10 percent between 2011 and 2022 (CARB, 2023a). As of the time of this writing, seventy-three national or regional carbon pricing initiatives have been implemented at various scales across the globe, covering approximately 23 percent of 2023 emissions (World Bank, 2023d).

Emissions trading has not been without problems. For one thing, carbon markets are less mature than those for other physical commodities, stocks, or bonds. Emissions credits are also attractive targets for theft, corruption, mismanagement, and abuse. In 2010 European emissions trading exchanges discovered that several participating governments, but particularly Hungary, were reissuing carbon credits, allowing the same emission allowance to be sold over and over again (Friesen, 2020; Reuters, 2010). This practice, if it became widespread, would eventually destroy the value of all credits held because the aggregate supply would not fall, allowing emissions to continue at their current rate indefinitely. In 2009 Europol discovered that cyber-criminals had absconded with nearly €5 billion ($5.68 billion) worth of carbon credits by breaking into the ETS computer network. The entire ETS system was temporarily shut down as a result. There was no clear assessment of liability, nor a plan to compensate credit-holders (Friesen, 2020; Europol, 2009). The structure and governance of carbon markets may improve in light of measures to strengthen Article 6 of the Paris Agreement during the COP 26 climate summit in Glasgow, UK, in 2021. Participants at COP 26 agreed in principle on mechanisms to trade carbon credits across borders, and to enable companies to reduce emissions in one country while taking credit for the reduction in a second country (UNFCCC, 2023).

Even if carbon markets eventually work as smoothly as trading corn futures, they can still become victims of their own success. California's LCFS program awards credits on a per- metric-ton of carbon basis if the carbon intensity of a fuel is less than the current state-mandated benchmark for that year. Fuels that exceed the benchmark, such as conventional gasoline or diesel, incur carbon deficits (CARB, 2023b). If the number of credits exceeds the number of deficits, a carbon bank holds the credits for subsequent drawdown in periods where the reverse holds true (more high carbon intensity fuels are being consumed than low carbon intensity fuels are being produced). Between 2011 and 2015 California's carbon bank steadily accumulated credits greatly in excess of the number of deficits being assessed against the use of conventional fuels (Sol Systems, 2022). Beginning in 2017, drawdowns on the bank have outpaced the production of new credits. This implies that fewer low carbon fuels, which would qualify for new credits, are being produced. Meanwhile, consumption of conventional fuels has also dropped in California. This appears to be a result of consumers beginning to embrace electric vehicles and other alternative transportation options. The combination of decreased fuel demand and the availability of ample bank credit supply has helped depress the price of carbon credits, which in turn decreases the incentive for producers to expand low carbon fuel operations. Ironically, the ensuing carbon credit market volatility has the potential to prolong production of conventional fuels.

THE SHELL GAME OF CLIMATE RISK TRANSFER

One of the most visible costs of climate change has been the transfer of risk to those who are less able to manage it. The increasingly fragile nature of home insurance markets is an example of this trend. Insurers have clearly determined that climate change is having a negative impact on their profitability (Hughes, 2023). The determination is supported by increasing amounts of data on the prevalence of natural disasters (UN, 2021). Despite the warning signs, in the United States more people are still moving to areas that are at a higher risk of climate-associated disasters than are relocating away from those locations (Cho, 2022). Part of the problem is that insurance companies use historical, rather than the latest, data to price policies. The Federal Emergency Management Agency (FEMA) has taken steps to modernize its flood insurance risk metrics for a national program that provides $1.3 trillion in coverage (Sherfinski, 2021), but flood insurance is typically optional outside of designated risk areas. A recent study by the First Street Foundation found that 14.6 million homes across the United States are at "substantial" risk of flooding, but only 5.9 million are in FEMA designated floodplain areas (Cho, 2022).

The rising frequencies of wildfires and floods has led insurance companies to cancel policies for homes in the worst affected locations (Dagher, 2023). In California, State Farm (correctly) predicted that its risk of fire claim liability was rising and—just months before the disastrous January 2025 Los Angeles wildfires—abruptly canceled approximately thirty thousand homeowners' policies in the affected areas, including ninety-five hundred in neighborhoods that burned (Eaglesham and Pulliam, 2025). As insurers withdraw, premiums rise or become unaffordable for a growing number of families. Widespread uninsurability will transfer a large number of single-family residential properties into the hands of big banks and investment firms, turning homeowners into tenants. If individual families lose their homes and cannot afford to rebuild, they will have no choice but to rent. Solutions to this problem are few and unappealing. Governments can use taxation or take on increased debt to rebuild homes, but this would amount to a publicly funded subsidy for those who *choose* to live in the path of natural disasters. The insurance industry could be more heavily regulated to compel coverage, but this would again lead to the socialization of risk as premiums would increase everywhere to protect the companies from insolvency. First Street (2025) estimates that as much as $1.47 trillion in U.S. home property value will be erased by 2055 due to climate change related factors and—in part—their effects on insurance markets.

The standard approach to managing property risk in the United States is still to rely on maps that categorize regions as more or less susceptible to various natural disasters. Historically, these maps have been more mature for flood than for fire or other risks. FEMA now publishes a National Risk Index that attempts to quantify risk potential for a variety of disasters, from avalanches, landslides, and ice storms to wildfires, volcanoes, and tsunamis. Flood insurance maps give at least some indication of the extent to which water historically encroached onto a particular piece of property. (This is the so called "one-hundred-year floodplain," meaning a flood would occur once in a hundred years). Since the frequency and severity of floods is changing in real time, relying on these maps is a less secure proposition than it once was. Wildfire risks are also multiplying at a rate that outpaces the information readily available to property owners and insurers. This makes mathematical risk metrics increasingly uncertain, driving up premiums as companies attempt to hedge against loss. The insurance industry's own models have quickly become outdated in the face of rapidly changing climatic conditions (Miettinen, 2021). Fire risk is particularly troublesome because the chance of accidents increases not only with population density, but also with people moving into forested areas. The risks to property can be interconnected: clearing ground cover to reduce fire danger may perversely make a region more prone to flooding or landslides. Vegetation helps to anchor soil in place and ash from fires can serve as a barrier to water absorption—allowing the water and loose soil to instead flow unimpeded downhill. A Pacific Palisades home that narrowly survived the January 2025 Los Angeles wildfires was subsequently lost to a landslide, likely because the home directly above it burned, resulting in a loss of vegetation and the deposit of ash on the nearby hillside (Richardson et al., 2025).

The management consultancy Deloitte (2023) surveyed insurers and their state regulators in 2019. Two thirds of regulators were not confident that insurers were prepared to manage the effects of climate change on the financial stability of their companies or the industry at large. The magnitude of potential economic loss, and the social and political dimensions of failures to manage the problem, make climate-driven insurability a necessarily public-private, collaborative issue. The concept of risk pooling will either be redefined to explicitly acknowledge the transference of risk to higher susceptibility popula-tions—those less able to shoulder the cost of rebuilding in a disaster-prone area—or a different market structure that can both compensate long-tenured homeowners and disincentivize unsustainable future development will need to emerge. A housing market that ignores climate risk at the point of sale and then asks FEMA to insure foresee-able losses after the fact is a form of taxation. Americans will eventu-ally tire of paying to build houses doomed to flood and burn.

We are likely to see people steadily migrate from coastal and arid regions that represent a great deal of economic value. Allowing that asset value to slowly (and literally) be submerged while turning a blind eye to those who suffer the loss of their homes is political cowardice. A rural neighborhood in Liberty County, Texas, serves as a cautionary tale for what can happen, even when honest efforts are made to manage retreat from flood-prone areas (Douglas, 2023). In a neighborhood where once as many as a hundred homes were occupied, no more than a dozen remain. Flooding along the Trinity River has intensified over the past ten years, and Hurricane Harvey delivered what turned out to be a lethal blow to the community in 2017. Intense rains have caused steady erosion of the riverbanks and surrounding flatlands. Roads are degraded to the point that emergency vehicles cannot navigate them, and municipal water service has been unavailable since 2020. After Harvey, FEMA made $6.1 million of funding available for property buyouts on a first-come, first-served basis. Purchased homes were demolished and the land transferred to the Texas Parks & Wildlife Department for a wildlife sanctuary (Bluebonnet News, 2020). When the money is gone, buyouts will cease. Those who miss out, or whose offers are insufficient to allow them to purchase homes elsewhere (there is no haggling allowed), are out of luck.

The seeds of another financial crisis lie dormant in this dilemma. Without concerted and costly efforts by the government to orchestrate an orderly retreat from economically uninhabitable areas, millions of people could lose everything. Too many folks have too much of their savings bound up in a single piece of residential property. Homeownership, like so many other things in this world, is proving to be too big to fail. A 2022 study by Columbia University's Climate School concluded that "[m]any states are facing a homeowners insurance market crisis as policies get more expensive and harder to come by. Florida's and Louisiana's homeowner insurance markets are particularly precarious" (Cho, 2022). The problem is of course not limited to the United States. Europe is also exceptionally vulnerable to drought and heat (Strovink, 2021), but has an adaptation advantage in terms of its wealth and commitment to state-sponsored social welfare policies. Communities in nations where public funding is insufficient to cover the costs of rebuilding or relocation face the prospect of large and potentially sudden increases in economic migration. If this migration occurs at a significant scale, it is likely to destabilize both source and destination countries in equal measure. Tensions over asylum seeking and economic migration to Europe and the United States in recent years already demonstrate the potential of this issue to become a political flashpoint. There is no wall high enough or ocean wide enough to deter destitute, thirsty, starving people from leaving their homes and finding their way to places where they can fight for the survival of their children. Paying people to relocate within their own countries or to otherwise adapt to disasters in their communities is certain to be more effective than trying to field armies large enough to hold the world's most unfortunate in tent cities along rich-country borders.

The same torrential rains, fires, floods, heatwaves, and ice storms that increasingly put residential property at risk also affect public infrastructure. Dams, bridges, roads, and electrical transmission lines have all been damaged by extreme weather. The resulting service outages can cost billions of dollars and place life and health at risk (Smith, 2022). Heatwaves in the Pacific Northwest in recent years have caused roads to buckle and electrical cables to melt (Frueh, 2021). As much as 25 percent of all critical infrastructure in the United States is at risk of failure due to flood alone, including two million miles of roadway that may become impassible more often than in the past (Kann and Nilsen, 2021). Worse yet, the rainfall records cities use to design flood control infrastructure is outdated, complicating plans to redesign and build additional infrastructure using funds from the American Rescue Plan Act or the Bipartisan Infrastructure Law. Some local officials worry that a federal project to update rainfall databases will not be complete until most or all the available funding has already been allocated—a situation that could lead to misallocation of resources (Sommer, 2023) and a false sense of security among residents of affected communities.

Corporations find creative ways to transfer climate risk while avoiding the costs of decarbonizing their operations. Divestiture is a tactic used by many large energy companies to increase the sustainability of their asset portfolios. Some of these firms are facing a great deal of pressure from climate-conscious communities or investors to make tangible progress towards meeting the Paris Agreement obligations of their home countries. For example, oil companies are selling their most heavily polluting wells to smaller, private producers (Meredith, 2022). It is perhaps obvious, but still worth mentioning, that new management does not necessarily imply superior environmental performance. In fact, smaller companies that take over marginally profitable, less desirable projects often have incentives to cut corners even more to generate financial returns. Private entities also face fewer obligations to report on the details of their operations than their publicly traded counterparts. Thus, new owners' negative impacts on future carbon emissions may go unnoticed.

Large mining companies are shedding risk by distancing themselves from plans to extract mineral rich nodules from the seafloor. They are selling their stakes in these ventures to smaller operators (Khan, 2023b). Seabed mineral extraction, which is essentially strip-mining of the oceans, risks greatly harming ecosystems (Schauenberg, 2023). The logic is simple: tiny organisms, which depend on minerals embedded in nodules at the bottom of the ocean, disappear, causing a chain reaction of nutrient loss up the aquatic food chain. Moreover, since much of the mining would take place in international waters, few if any regulations would apply or be enforceable. The incentives for companies to cause untold pollution and damage to reduce what would otherwise be exorbitant costs will be overwhelming. Yet, these nodules represent a sizable quantity of the raw materials needed to implement the electrification of industry and transport. The principal tradeoff we face, as usual, is one of cost versus convenience. Will people be willing to pay a high enough premium for these minerals to guarantee that they are harvested in ways that do not destroy the world's oceans? Or will deep sea mining become the source of another unintended, but foreseeable, disaster?

A sustainable energy transition cannot succeed if it is envisioned primarily as a way for people to make money. Environmental pollution exists for the very reason that it is always cheaper to dump waste somewhere out of sight than to pay for reducing it or mitigating its negative impacts at the outset. Restated from a thermodynamic perspective, the most irreversible (entropy generating) processes are those that occur most spontaneously. Energy and carbon markets are, and will continue to be, useful ways to allocate resources. What they should *not* be seen as are ways to "beat the system" and somehow save the world while getting rich. Measuring and acknowledging the true costs of living in a low carbon economy requires a brand of humility and acceptance that is hard to come by in today's world of boundless technological optimism and quarterly earnings reports. Resources are limited, and choices have consequences. Policymakers owe it to their constituents to be honest enough to admit to them that they can't have it all. With that in mind, I will close this discussion with a look at the kinds of climate programs that are most likely to actually be implemented under the political and economic conditions we find ourselves in today.

CHAPTER 9
A ROAD PAVED WITH GREEN INTENTIONS

Climate scientists can warn humanity of dangers from unabated greenhouse gas emissions. Engineers can develop technologies to remove or slow the rate of those emissions. Economists can predict the impact of climate change on productivity and the wealth of nations. Social scientists can articulate the impact of climate change on culture and politics. The financial sector can maintain markets for carbon, power, minerals, and other vital energy commodities. Governments can design the climate and energy policies, trade agreements, and regulatory actions needed to launch, nurture, and enforce a clean energy transition. Yet, even when all these activities are synchronized in perfect harmony, it is citizens and consumers, individually and collectively, who must wield their power to transition to a just and sustainable future.

At the time of this writing, the citizens of Earth have manifestly declined to do so. Some people buy electric cars; others go vegan; still others chain themselves to the gates of coal mines or deface artworks in museums. As for myself, I try to use as little gasoline as possible; I fly less; I buy fewer toys. But I am not, nor is anyone else, living a carbon free existence. The election of Donald Trump to a second presidential term has given many on the left a convenient excuse for our collective failure to halt greenhouse gas pollution. Trump is certainly no friend of the environment, and he makes a good cartoon villain, but the future of the global climate is not his responsibility alone; we all share that burden. The left blames "climate change deniers" while continuing to live their ordinary consumption-centric lives and vigorously fighting against infrastructure being built anywhere near their homes. The right, for their part, need to be realistic. They are foolish to ignore the economic and social calamities that are already beginning to accompany climate change. Maybe that will change after a few more Category 5 hurricanes decimate major cities. Time will tell. And climate change is not a capital allocation problem in the sense of neoliberal-style free trade. We will not be spending our way out of this mess.

There is no one right way to address human induced climate change. The scope, shape, and pace of government action in every nation must match the capabilities and needs of its citizens. This effort requires being mindful of the unique characteristics of the places in which they live, the specific effects of climate change on their homes and livelihoods, and realistic assessments of the resources that will be available to them. This inevitable ambiguity is, however, no excuse for world leaders dissembling or offering false assurances. I am reminded of a phrase that came into vogue around the turn of the twenty-first century: "fail fast." Startup companies, particularly those developing software, found it useful to work quickly on potential solutions to business problems. They were willing to abruptly change course— even in the face of enormous sunk costs—if the solution ultimately showed itself to be unviable. Engineers often refer to this approach to design as "cut and try." It is the antithesis of the time-honored admonishment to "measure twice, cut once." Electric vehicles come to mind here: They aren't meeting the moment. They are too expensive, and the charging infrastructure needed to make them even marginally feasible for use throughout a huge road trip-crazy nation like the United States is not forthcoming. On January 6, 2025, the Trump Administration ordered U.S. transportation officials to halt federal funding to the states for construction of electric vehicle charging stations (Hiller, 2025b). But to be honest, in most parts of the country, consumers aren't buying electric cars anyway. In December 2024 electric vehicles accounted for "a new record" of 8.8 percent of new car sales (Johnson,

2025). This is not good enough and, given how much money has already been spent trying, it isn't likely to be soon.

If we can't agree to go all-in on high-speed rail the way Europe, China, and Japan have, then we will need to take a second look at hydrogen for transportation. As we've seen, the use of hydrogen as an energy carrier can take many forms. Given that slightly over half of American voters support our current fossil fuel-friendly administration, we would be well advised to reach out to the oil and gas companies for assistance. I think e-fuels are probably a compromise that conservatives might rally behind. They are not a silver bullet solution to the problem of climate change, but they would represent an improvement over our current failures to make progress. We should also work with the hydrocarbon industry to build dual-fuel gas power generation infrastructure: plants, pipelines, compressors, and storage facilities. We can use natural gas for as long as we need (want?) to and have a way to gracefully transition to hydrogen—blue, green, pink (made using nuclear energy)—once we finally decide we can't postpone it any longer.

There is a lot to be said for distributing resources broadly to maximize our chances of discovering the most innovative and efficient ways to decarbonize the global economy. Failing fast may winnow the field of options, but only if we have the right objective in mind. There, of course, lies the rub. Are we hoping to use markets to build new energy businesses that maximize returns for investors? Are we hoping to minimize humanity's contribution to the rise in global concentrations of greenhouse gases? Are we hoping to reduce the rate of global average temperature increase, irrespective of greenhouse gas emissions? Are we doing these things for ourselves, or for people who may be born centuries from now? I used to be a product manager, and aspiring product managers are told always to keep asking the same question of prospective customers: what problem are you trying to solve? You can't provide solutions until you understand the problems. A great concern I have about climate change mitigation is that no consensus agreement exists regarding what precisely it is we are trying to accomplish. Perhaps nothing exemplifies this confusion more than the electric car industry. Are we building electric cars to reduce our carbon footprint, to colonize other planets, to eliminate human drivers, or simply to create a new lucrative consumer product market? The "invisible hand" of the free market cannot solve global environmental problems, and the way we have structured markets in the United States is especially unsuitable for that task. We have chosen to give a piece of paper all the rights of a living, breathing human being. Yet these corporate "persons" have no conscience—only a single-minded focus on increasing their own size measured in dollars. Corporations cannot be expected to consider and promote concepts such as intergenerational justice or the preservation of ecosystems. Only *we* can do that, and it is *not* profitable in the sense of a quarterly earnings report.

But that is the system we have to work with, and—bearing that in mind—we also cannot throw an entire generation of retirees under a bus by abruptly adopting an unbuffered degrowth policy: redistribution is an absolute must. Taxation is a controversial topic under any circumstances, and convincing people that they should pay more for clean energy is a tough sell. Convincing them that they should pay more for foreigners to have clean energy is even tougher. Fortunately, there is a small group of people who can afford to lend a hand. In 2020 the world's 2,153 billionaires held as much wealth as the poorest 60 percent (4.6 billion people) of the population. The twenty-two richest men in the world are worth more than all the women in Africa combined (Picchi, 2020) and the concentration of global wealth is only accelerating. If governments cannot bring themselves to tax their richest citizens at a level sufficient to fund their most urgent priorities, then perhaps they have lost a certain amount of legitimacy. On the other hand, the tension between the desire of elites to insulate themselves from social obligations and the need for governments to draw upon their wealth is ancient. In feudal times, monarchs required landed knights to provide either cash or military service in times of national emergencies. In England, the king's perceived (and real) abuse of this power of taxation led to the drafting of the Magna Carta (McKechnie, 2021), from which Americans derive our famous demand "no taxation without representation." Nevertheless, for a project as sweeping in scope and complexity as the complete and unprecedented transformation of the entire world's systems of energy production, delivery, and consumption, we will have to come to an understanding. To use a line attributed to the legendary Depression-era bank robber Willie Sutton, we will have to rely in no small part on the rich to finance a clean energy transition because "that's where the money is."

Much of today's wealth has been accumulated from the growth of corporate stockholdings. It is worth remembering that a great deal of that growth has been heavily subsidized over the years by public funds in one way or another. These subsidies were collective investments on which we can all justifiably demand a return. The International Monetary Fund estimates global fossil fuel subsidies to have totaled $7 trillion in 2022 (7.1 percent of global GDP). They are expected to increase to 8.2 percent of global GDP by 2030 (IMF, 2025; Nemetz, 2022). Removing these subsidies to expose the true market price of carbon could lead to a 36 percent reduction in emissions and raise governmental revenue by 3.8 percent of global GDP. The message is as simple as any can be: we can continue *paying* people to cause carbon pollution, or we can start *charging* people for the privilege of doing so.

This does not eliminate the responsibility of the rest of us to do what we can with what we have. Americans, for example, have been told for generations that we are exceptional, that we can reach our dreams—whatever they may be—if only we are willing to work hard enough. This admirable, rags to riches ethic of rugged individualism is romantic, and may even be useful as encouragement. On the other hand, the American Dream has given too many people the impression that they can have everything they want, whenever they want it, at an affordable price, without any consequences. What I have tried to explain in this book is that that is not a realistic way to think about the world. Every energy design choice we make will constrain us in some way. Those constraints will not necessarily be limited to technological factors. They will undoubtedly involve social, political, environmental, economic, and cultural factors as well. It is imperative that each of us embraces this fundamental truth when considering our consumption decisions and when evaluating claims made by those who offer what appear to be magical solutions to energy challenges. Whatever we are trying to accomplish, one thing is certain: it does not matter in the least whose "fault" it is that we are in this predicament. Friesen (2020) presents a metaphor of carbon pollution as the steady filling of a lake with water from pipes, each pipe owned by a single nation. It is undeniable that Europe and the United States were responsible for the largest *initial* quantities of carbon emissions, flowing fast through their pipes as the West grew wealthy during the Industrial Revolution. It is equally undeniable that China is now responsible for the largest *current* level of emissions as it builds at breakneck speed to earn the same standard of living their neighbors have long enjoyed. Who can blame them? And yet, we must make *future* changes to our energy systems based on *today's* reality. Meanwhile, billions of people in Asia, Africa, and Latin America are waiting their turn to join the party. These people have neither the wherewithal to take an express elevator to prosperity nor the means to avoid the worst consequences of climate change: they are trapped in the middle. Their goals are crystal clear: they want more energy and less environmental collapse. Many of the poorest nations are largely relegated to waiting for us to do something that can make this happen for them.

There are other things we can do. One of the first and best is to complete the work already started to establish global standards and markets for carbon trading. The latest global meeting of signatories to the Paris Agreement (COP 29 held in Baku, Azerbaijan on November 12, 2024) resulted in an agreement to finalize international carbon market standards that had been left as provisional since the 2015 Paris meeting (UNFCC, 2024). The most suitable locations for carbon storage are not distributed in equal proportions among the nations, nor are they distributed in proportion to expected future emission levels. Cross-border trading of carbon is, therefore, an essential vehicle for making optimal use of storage resources. The European Parliament's move to impose a border carbon tax (Dalton, 2023a) is a hopeful sign. The law would create a tariff on goods tied directly to measurements of the emissions associated with their manufacture and delivery. Compliance would spur more accurate and sophisticated carbon accounting in almost every sizeable corporation. Competition to reduce tariff costs would help establish globally consistent prices for different classes of carbon offsets. These would include embedded carbon storage (for example, in forests, oyster beds, or kelp farms); geological carbon sequestration; and carbon recycling programs. Price certainty would help drive greater adoption of those technologies. How much adoption is still an open question. The net effect on atmospheric carbon concentrations will be decided in a race between mitigation efforts and the unceasing global demand for high-energy lifestyles. When I worked at ExxonMobil, management was fond of pointing out that gasoline was cheaper per gallon than any liquid you could buy at the grocery store. It was true, and it is emblematic of the problem we face. If carbon becomes expensive, people will use less of it. If not, it is difficult to imagine a low-carbon economy taking root. Without a bona fide, enforceable carbon tax, the clean energy transition will almost certainly falter.

Americans perennially attempt to diet by eating the same amount of food, while choosing something different to eat each time around (low carb, paleo, keto, gluten free). There will be a great temptation to simply build our way out of the climate crisis, consuming just as many resources, only different ones than before. It behooves us to remember three things. First, the law of unintended consequences is alive and well. Second, one can never overestimate the sense of entitlement among a group of people who have become accustomed to having plenty of something. Third, humans are imbued with an innate fight or flight response that only engages once circumstances become dire. Like the proverbial frog that allows itself to be cooked to death slowly in a pot of very gradually boiling water, wealthy societies have so far been unfazed by incremental climatic changes because they have not experienced sharp and immediate pain. Even as late as the turn of the twenty-first century, most tangible effects of climate change were largely undetectable by the public. This convenient situation has allowed oil and gas companies to vigorously deny that a climate crisis exists while scrambling for strategies that will allow them to profitably adapt to it.

GEOENGINEERING—WHEN ALL ELSE FAILS

If we can't—or won't—live within a sustainable carbon budget, we can try instead to blunt the harmful effects of emissions. Geoengineering refers to planetary-scale modifications of the environment designed to disrupt and control the natural world. If this sounds extreme, that's because it is. Ideas floated to combat the effects of greenhouse gas emissions include: injecting huge numbers of reflective particles into the upper atmosphere; seeding low altitude clouds with particles to "brighten" them; modifying the reflective properties of high altitude clouds to increase the atmosphere's transparency to outgoing radiation; fertilizing the oceans to grow legions of carbon-absorbing microorganisms; and placing large numbers of reflective panels into orbit around the Earth to block incoming sunlight (NASEM, 2021; Keith et al., 2020; Babakhani et al., 2022). All the above can be classified as "solar geoengineering" because they attempt to moderate the amount of solar radiation reaching, or being absorbed by, the Earth. In truth, we have been engineering the planet for over a century in the form of atmospheric nitrogen fixation (recall the Haber-Bosch process). This project is utterly inseparable from modern agriculture, because it is the only method we have to supply enough fertilizer to feed our current population. Direct air capture of carbon is also an example of geoengineering that, as we have discussed, has many proponents—especially among oil and gas industry veterans. Some see geoengineering as a chance to rewind history without consequences. This view is demonstrably false. Every one of the above technologies adds additional complexity to the climate, if not by directly changing the chemistry of the air, land, or water, then by allowing for atmospheric carbon concentration to continue to increase while unevenly reducing terrestrial temperatures and prompting regional changes to the water cycle.

The saturation of Earth's atmosphere with carbon since 1750, and the consequent alteration of the climate, is itself a natural experiment in geoengineering gone awry. There is currently no broad consensus on the wisdom or efficacy of any prospective future geoengineering strategy that has been envisioned to date., Many experts urge caution and restricting such technologies to small-scale experiments unless we can gain more confidence in their consequences (Union of Concerned Scientists, 2020; Nature, 2021b; Chu, 2020; Boyd and Vivian, 2019; NASEM, 2021). Geoengineering does not actually address the underlying problem: emission of greenhouse gases. If we were ever to stop any geoengineering program, we could face a nearly instantaneous (and perhaps even larger) rate of global warming, a phenomenon referred to as a "termination shock" (Morton, 2015). Nevertheless, geoengineering will probably prove irresistible to politicians. It is the type of solution that readily lends itself to being used as a golden hammer. Abraham Maslow famously remarked that, if the only tool you have is a hammer, every problem looks like a nail. Applied liberally, geoengineering may allow governments to, on one hand, appear to be sparing no expense to tackle an existential problem, and, on the other hand, sweep away politically inconvenient complexities associated with other climate mitigation strategies such as mandatory energy conservation, universal electrification, or infrastructure switching.

The allure of geoengineering, which proposes to relieve the pain of climate change while allowing business to proceed largely as usual, will only grow stronger the longer we avoid eliminating the root cause of human-induced warming—emitting more carbon into the atmosphere. Remember that it is not the *rate* of emissions, nor the *carbon intensity* of operations, but the accumulated physical mass of carbon floating above us that influences climate. Even if we stopped emitting carbon tomorrow, as much as a quarter of the three hundred billion metric tons added to the atmosphere since 1945—the point at which consumption of everything shot up like a rocket—could remain there for thousands of years (McNeill and Engelke, 2016).

The incentives surrounding geoengineering, even where it is optimally designed, are misaligned. There is no way to prevent any one nation from making its own decisions about how and at what scale to deploy geoengineering technologies. Treaties can be negotiated, but any OPEC member can comment on how well oil quotas are controlled if one nation benefits more by increasing production in defiance of the others. If a nation is not seeing the results it would like from geoengineering, it will find a way to accelerate, or deploy countermeasures against, the project. A particularly alarming scenario would be the prospect of induced failure of seasonal Indian monsoons that could result from unanticipated changes to regional ocean current, air current, and rainfall patterns. It is possible that such a catastrophe could be attributable to a geoengineering project that causes changes in land-sea temperature gradients. Such a fiasco would leave a nuclear-armed nation of over one billion people facing famine and potential political collapse (Morton, 2015). A second misaligned incentive is the temptation to profit from business as usual by gradually increasing geoengineering efforts in lockstep with steadily increasing carbon emissions. Project creep is especially likely if the geoengineering systems turn out to be cheaper and easier to implement than austerity, electrification, or other decarbonization measures (and they almost certainly will be—at least in the short term). It is incredibly difficult to tell voters or shareholders that, although a solution to a problem exists, it will not be used because a better, less convenient, more expensive one really should be implemented instead.

The pinch points of any clean energy transition will largely be materials related. That realization implies a reduction in consumption of just about everything. A nation's adversaries would likely turn radical, unilateral economic degrowth to their advantage. So, the best chance we have to arrest the worst impacts of climate change is to wind down our most wasteful and unnecessary habits over a few generations using wealth redistribution—aggressive taxation of the ultrarich—to soften the blow. The sudden and massive availability of cheap energy after 1945 enabled a several orders of magnitude increase in deforestation, urban sprawl, agricultural production, and manufacturing. Our unbridled growth has consumed resources at an unsustainable rate (McNeill and Engelke, 2014) that could probably never be repeated. The fact that the energy used was (is) fossil in nature is not germane to the rate of associated resource extraction, per se. In fact, if a hypothetical new source of energy were to emerge that was even cheaper, more widely available, and freer of environmental consequences than what we have now, it would only serve to accelerate every other form of extractive activity. This is—once again—a manifestation of the Jevons paradox: the more cheaply a resource *can* be utilized, the more of it *will* be utilized.

Politicians, experts, and "influencers" need to be brutally honest in this regard. There is no scenario in which humanity "solves" climate change and then goes back to living as Americans do today. We will have to learn to live simpler lives—somewhere between a prehistoric existence and a twenty-first century Western middle-class lifestyle. Where exactly along that continuum we land depends crucially on the size of the stable global population. To date, the most likely scenario appears to be that everyone who cares at all about reducing carbon pollution will choose one or two core things to focus on: driving less; flying less; eating less meat. All of it together will not be enough, and that's the hard truth. As the ensuing cost to society from storms, floods, fires, increasing food prices, and immigration becomes too large to ignore, a collection of emergency geoengineering projects will undoubtedly emerge. The cure might just work, after a fashion, and for a while. Its unintended consequences will be managed as well as possible. The usual suspects (war, plague, political intrigue) will, as always, hamper any coordinated, worldwide program of engineered sustainability. But we have to try, because we will otherwise run out of metals, oil, and the ability to manage an inhospitable environment.

Wishful thinking and history have led many of us to believe that a Green New Deal can lift us out of this crisis. The Roosevelt Administration worked wonders for the United States during the Great Depression. The Tennessee Valley Authority, corresponding flood control projects around the Mississippi Delta, and damming of the Columbia River in the Pacific Northwest supercharged the nation's economy and alleviated poverty for millions. That energy transition of yesteryear was possibly also the single most important factor contributing to our victory in World War II. Hydropower built our Air Force by enabling us to run giant aluminum smelters day and night. It further enabled the Manhattan Project to succeed (Palumbi, 2023) by energizing the production of plutonium at the Hanford Site in Washington. Our world was dealt a winning hand of abundant natural capital, but we have overplayed it. Thermodynamic scarcity is destined to make each new exploitation of our remaining resources less impactful and more expensive than the last. The hydrocarbon industry knows this. They have been fighting the demon of peak oil for decades by exploring increasingly remote, marginal, and ecologically fragile regions. We cannot sustainably *develop* our way to a low carbon future because diminishing returns to future energy investments will increasingly handicap our ability to meet society's other manifold needs.

Environmental sustainability has been aptly described as an exercise in asset management (Nemetz, 2022). Earth is a closed system (neglecting the mass we occasionally gain from collisions with meteorites, lose through launching satellites, and other tiny transfers of matter). The stock of natural capital we have available to work with is, for all intents and purposes, constant. By applying ingenuity (human capital), we transform natural capital into manufactured capital. It can never be perfectly restored to its original form because its entropy increases. A greenfield project site developed into a factory might eventually be revived to look like a green field, but it will never again be *the same* green field. Exploitation of increasingly lower quality—and more thermodynamically scarce—resources is inevitably accompanied by greater entropy production, and entropy ends up becoming a form of social debt. To the extent that the costs of our entropic waste are paid by someone else, somewhere else (if they are even comprehended at all) the incentive to change our own behavior is lacking.

In that respect lies another big risk of geoengineering. Even at their best, geoengineering projects steadily grow into social debts. Governments will make minimum payments toward these debts, but the principal balances will continue to grow as technology depreciates and competing claims emerge to deflect money and urgency. The best working example of this debt cycle is humanity's utter dependence on synthetic fertilizers. Global agriculture would cease to function overnight in the absence of ammonia-derived fertilizers manufactured courtesy of the Haber-Bosch process. Our geoengineering of the planet's nitrogen cycle in this way has become a more essential entitlement program than any pension or healthcare program could ever be. Direct air capture of carbon dioxide, a floating array of low orbit mirrors, or a perpetual aerosol spraying campaign would be nothing but additional thermodynamic credit cards offered to us at deceptively low introductory rates. The human race can't afford to declare bankruptcy and start over if geoengineering becomes too great a burden to continue and our descendants surrender to accelerated climate change. No one wants this to happen, so we—here and now—need to live within our material, energy, and political means instead.

Gradual depopulation from declining birthrates, coupled with the best achievable future technology, can enable humanity to achieve a long-term equilibrium energy budget at some livable rate of carbon emissions. Precisely which goods and services we consume will be different, of course. What seems inevitable is that the world will have to slow down tremendously: We will have much less mobility than we have become accustomed to over the past seventy-five years. Rampant consumerism will have to give way to an ethos of relative frugality that will challenge our social contracts. Whether we choose to make common cause with one another and construct an egalitarian and less wasteful world, or whether the many will be exploited by the few, is the same choice that we make every day. We cannot expect to continue living in post-World War II *les trente glorieuses* on a planet inhabited by eight to ten billion people. If this concerns you, then I urge you to begin finding every way you can to consume less right away. Much of what we buy is so unnecessary that we don't even remember that we own it after a few weeks or months. Look in any American garage. You probably won't find a car, but rather enough abandoned sporting goods to supply the Olympics, broken appliances, a dusty and lonely workbench, and cans of dried out paint. Marie Kondo, who became famous as an advocate of decluttering, notes wisely that, if you don't even remember that you possess something, it can't be adding much joy to your life. Travel, too, can be simpler, yet still rich and fulfilling. Gradual and thoughtful reduction of our consumption, while easing the economic transition for those who are nearer to the end of their life's journey, is preferable to ignoring the day of reckoning until it becomes a jarring catastrophe for everyone all at once.

At this point I have to say one more thing about "AI." First and foremost, it is not intelligence. It is a mathematical tool that enables a different flavor of computer programming from what scientists and engineers have been reliant on for the past few generations. When I was in college we used to refer to primitive forms of what people today call "machine learning" as "curve fitting." One can draw the best fit straight line through a set of points plotted on a graph and use the equation of that line to estimate values of the corresponding dependent variable that are not represented on the graph by actual, measured data points. This technique is known as linear regression. Neural networks do much the same thing, but in billions of dimensions all at once. The values ascribed to individual neurons in a deep network are adjustable parameters that function analogously to the coefficients in a regression equation. These networks are very useful. They are not, however, going to save the world. Machine learning techniques are very helpful where one has not developed the requisite theoretical understanding of a system to use phenomenological, or "physics-based" models. In some cases, we have no other choice, but we should never be lulled into the sense that we have somehow "beaten Nature at her own game."

Why am I talking about this? Nuclear electricity generation has the potential to relieve a great deal of our current climate and energy security stresses, and to do so for quite some time—probably centuries. It, too, is not a silver bullet solution to climate change, but it is a very important tool we have access to right now. Of course, we must be willing to use it. That means we have to accept the reality of long-term nuclear waste disposal. We also must be more willing to consider recycling and reprocessing nuclear fuel. The French are on board with this, planning to expand their already impressive fleet of civilian reactors (Schechner and Fitch, 2025). The problem is in their motivation: they want to use all their new nuclear power to achieve supremacy in AI. Here in the United States, similar efforts are underway, including a plan to restart the long-mothballed Three Mile Island nuclear plant so it can sell electricity exclusively to Microsoft for powering data centers (Valinsky, 2024). We are still burning plenty of gas and coal. We are even keeping some of our oldest fossil fueled power plants open long past their planned retirement dates because of our immense unmet electricity demand (Gaffney and Rojanasakul, 2025). And our priorities? Self-driving cars, chatbots to write term papers for lazy students, and vanity cryptocurrencies whose values are tied to the popularity of internet memes (Snider, 2025). Why can't we instead focus on first providing everyone with enough electricity for home heating and cooling, clean water, lighting, and reliable communications to an extent that does not put the global environment at risk?

Human nature seems to be biased against such prudence. Perhaps we will take one more giant leap for mankind, betting the future of our species on the chance to live as interplanetary hunter-gatherers and devouring the resources of new worlds with the same voracity as we've done on this one. It's a recurring theme: the people of Easter Island; the Anasazi of the American Southwest; even the march of Manifest Destiny that fleshed out the modern borders of the United States—we consume everything around us and then move somewhere else to consume more. So, some say, why not the moon, Mars, and untold worlds beyond? To me, the failure to sustain life on Earth would be a profound articulation of our foolishness. Moreover, while science fiction can be entertaining, it is not likely that anyone alive as I write these words will migrate to Mars. When we make plans at scales that can spell life or death for civilizations or entire species, we should have the humility to pause and consider our limitations as well as our aspirations. Our endowment of exergy—our ability to extract useful work from the environment around us—is finite, whether embodied in raw materials or in concentrated sources of energy. Each action we take, each thing we create, is ultimately irreversible: we can't recover spent natural resources any more easily than we can unbake a cake. Museums around the world hold vast collections of artifacts from human civilizations that rose and fell over thousands of years. Almost all those people used virtually identical amounts of energy in a global economy that hardly grew the entire time. In the blink of an eye, over-consumption has now threatened to end it all. We will need to start making wiser choices about how we use energy and materials if we hope to have humans, or whatever they evolve into, roaming this planet thousands of years from now.

BIBLIOGRAPHY

Abercrombie, R. E. et al. 2021. Does Earthquake Stress Drop Increase With Depth in the Crust?, *JGR Solid Earth*, 126(10), e2021JB022314, https://agupubs.onlinelibrary.wiley.com/doi/10.1029/2021JB022314.

Albert, Nico. 2018. Food, Culture, and Storytelling, *Food*, Public Broadcasting Service, https://www.pbs.org/food/stories/food-culture-storytelling.

Al Ghussain, Alia. 2020. The biggest problem with carbon offsetting is that it doesn't really work, *Greenpeace UK*, https://www.greenpeace.org.uk/news/the-biggest-problem-with-carbon-offsetting-is-that-it-doesnt-really-work/.

Alhujaili, Asmaa et al. 2023. Insects as Food: Consumers' Acceptance and Marketing, *Foods*, 12(4), 886, https://pmc.ncbi.nlm.nih.gov/articles/PMC9956212/.

Allied Market Research. 2021. *Carbon Capture, Utilization, and Storage (CCUS) Market by Service (Capture, Transportation, Utilization, and Storage), Technology (Pre-Combustion Capture, Oxy-Fuel Combustion Capture, and Post-Combustion Capture), and End-Use Industry (Oil & Gas, Power Generation, Iron & Steel, Chemical & Petrochemical, Cement, and Others): Global Opportunity Analysis and Industry Forecast, 2021-2030*, Report Code A12116, https://www.alliedmarketresearch.com/carbon-capture-and-utilization-market-A12116.

American Carbon Registry. 2023. https://americancarbonregistry.org/.

American Society of Civil Engineers. 2023. *Road Infrastructure | ASCE's 2021 Infrastructure Report Card*, https://infrastructurereportcard.org/cat-item/roads-infrastructure/.

Anchondo, Carlos et al. 2023. EPA says carbon capture is within reach. Utilities aren't biting. *E&E News EnergyWire*, https://www.eenews.net/articles/epa-says-carbon-capture-is-within-reach-utilities-arent-biting/.

Anderson, Jared. 2022. Wind, solar resources should be removed from PJM capacity market: P3 trade group, *S&P Global Commodity Insights*, https://www.spglobal.com/commodityinsights/en/market-insights/latest-news/energy-transition/020222-wind-solar-resources-should-be-removed-from-pjm-capacity-market-p3-trade-group.

Arbor Day Foundation. 2023. *Loblolly Pine*, https://www.arborday.org/trees/treeGuide/TreeDetail.cfm?ItemID=899.

Arezki, Rabah and Per Magnus Nysveen. 2023. *End of the Line*, International Monetary Fund, https://www.imf.org/external/pubs/ft/fandd/2021/06/the-future-of-oil-arezki-and-nysveen.htm.

Argonne National Laboratory. 2020. *Unraveling the Mechanisms Responsible for Hydrogen Embrittlement of Steel*, https://www.aps.anl.gov/APS-Science-Highlight/2020-06-02/unraveling-the-mechanisms-responsible-for-hydrogen-embrittlement.

Associated Press. 2023. *Norway protests target wind farm on land used by herders*, https://apnews.com/article/norway-wind-turbines-sami-herders-protest-c97371074db742c0654da522c5a9f379.

Associated Press. 2022. *Germany's Merkel defends decision to get Russian natural gas,* https://apnews.com/article/russia-ukraine-business-germany-lisbon-2a8722c63c7ba84923c12eb0e1f8d35c.

Babakhani, Peyman et al. 2022. Potential use of engineered nanoparticles in ocean fertilization for large-scale atmospheric carbon dioxide removal, *Nature Nanotechnology*, 17:1342-1351, https://www.nature.com/articles/s41565-022-01226-w.

Badger, Emily and Alan Blinder. 2023. How Air-Conditioning Conquered America (Even the Pacific Northwest), *New York Times*, https://www.nytimes.com/2017/08/04/upshot/the-all-conquering-air-conditioner.html.

Baer-Sinnott, Sara. 2023. Food Is a Window to Cultural Diversity, *U.S. News & World Report*, https://health.usnews.com/health-news/blogs/eat-run/articles/food-is-a-window-to-cultural-diversity.

Bakke, Gretchen. 2016. *The Grid: The Fraying Wires Between Americans and Our Energy Future*, New York: Bloomsbury Publishing.

Bakker, Conny et al. 2019. Circular Product Design: Addressing Critical Materials through Design. in *Critical Materials: Underlying Causes and Sustainable Mitigation Strategies*, S. Erik Offerman, Ed., Singapore: World Scientific.

Bandiera, Luca and Vasileios Tsiropoulos. 2019. A Framework to Assess Debt Sustainability and Fiscal Risks under the Belt and Road Initiative, *World Bank*, https://documents1.worldbank.org/curated/en/723671560782662349/pdf/A-Framework-to-Assess-Debt-Sustainability-and-Fiscal-Risks-under-the-Belt-and-Road-Initiative.pdf.

Barrett, Joe. 2023, On One Texas River, Four Dam Failures Show Harsh Reality of Aging Infrastructure, *Wall Street Journal*, https://www.wsj.com/articles/on-one-texas-river-four-dam-failures-show-harsh-reality-of-aging-infrastructure-bfc0f279.

Bartlett, Jeff S. and Ben Preston. 2023. Automakers Are Adding Electric Vehicles to Their Lineups. Here's What's Coming., *Consumer Reports*, https://www.consumerreports.org/cars/hybrids-evs/why-electric-cars-may-soon-flood-the-us-market-a9006292675/.

Baurick, Tristan. 2024. Surprise bids revive hope for offshore wind in Gulf of Mexico after feds cancel lease sale, *Associated Press*, https://apnews.com/us-news/gulf-of-mexico-coastlines-and-beaches-chicago-general-news-f5250c5ad324b4a27fd6139ab4e3b37c.

BBC. 2023. *The formation and usage of fossil fuels*, https://www.bbc.co.uk/bitesize/guides/zxxq4xs/revision/1.

Beaumont, Thomas and McFetridge, Scott. 2021. Red meat politics: GOP turns culture war into a food fight, *Associated Press*, https://apnews.com/article/business-cultures-climate-change-environment-and-nature-government-and-politics-fb2a32e232053326f6f1ebfaadcc2e40.

Bekdemir, Cemil et al. 2022. H2-ICE Technology Options of the Present and the Near Future, *SAE Technical Paper 2022-01-0472*, https://www.sae.org/publications/technical-papers/content/2022-01-0472/.

Benoit, Bertrand. 2025. Meet the Blue-Collar Voters Making Germany's AfD Mainstream, *Wall Street Journal*, https://www.wsj.com/world/europe/meet-the-blue-collar-voters-making-germanys-afd-mainstream-79350785.

Benoit, Bertrand et al. 2024. The Progressive Moment in Global Politics Is Over, *Wall Street Journal*, https://www.wsj.com/world/global-politics-conservative-right-shift-ea0e8d05.

Besser, Linton. 2021. Dead white man's clothes, *Australian Broadcasting Corporation*, https://www.abc.net.au/news/2021-08-12/fast-fashion-turning-parts-ghana-into-toxic-landfill/100358702.

Beswick, Rebecca R. et al. 2021. Does the Green Hydrogen Economy Have a Water Problem?, *ACS Energy Letters*, 6(9):3167-3169, https://pubs.acs.org/doi/10.1021/acsenergylett.1c01375.

Bettenhausen, Craig. 2021. The life-or-death race to improve carbon capture, *Chemical & Engineering News*, https://cen.acs.org/environment/greenhouse-gases/capture-flue-gas-co2-emissions/99/i26.

Binkley, Collin. 2023. Jaded about education, more Americans are skipping college, *Los Angeles Times*, https://www.latimes.com/world-nation/story/2023-03-09/more-americans-are-skipping-college.

Birnbaum, Michael. 2023. Europe needs energy. Moroccan solar may be a clean solution., *Washington Post*, https://www.washingtonpost.com/climate-solutions/2023/04/13/morocco-europe-solar-desert/.

Bishop, Breanna. 2022. Lawrence Livermore National Laboratory achieves fusion ignition, *Lawrence Livermore National Laboratory*, https://www.llnl.gov/news/lawrence-livermore-national-laboratory-achieves-fusion-ignition.

Bittle, Jake. 2023. *The Great Displacement: Climate Change and the Next American Migration*, New York: Simon & Schuster.

Bloomberg NEF. 2023a. *A Power Grid Long Enough to Reach the Sun Is Key to the Climate Fight*, https://about.bnef.com/blog/a-power-grid-long-enough-to-reach-the-sun-is-key-to-the-climate-fight/.

Bloomberg NEF. 2023b. *Global Net Zero Will Require $21 Trillion Investment In Power Grids*, https://about.bnef.com/blog/global-net-zero-will-require-21-trillion-investment-in-power-grids/.

Bloomberg NEF. 2023c. *New Energy Outlook: Europe*, https://about.bnef.com/new-energy-outlook-series/.

Bloomberg NEF. 2023d. *New Energy Outlook: Grids*, https://about.bnef.com/new-energy-outlook-series/.

Bluebonnet News. 2020. *Buyouts still underway for flood-prone communities in Liberty County*, https://bluebonnetnews.com/2020/01/02/buyouts-still-underway-for-flood-prone-communities-in-liberty-county/.

Blunt, Katherine. 2022. *California Burning: The Fall of Pacific Gas and Electric—And What It Means for America's Power Grid*, New York: Penguin Publishing Group.

Booth, William. 2023. In Scotland, making whisky with energy from wind, wood chips and tides, *Washington Post*, https://www.washingtonpost.com/climate-solutions/2023/03/21/scotland-making-whiskey-with-energy-wind-wood-chips-tides/.

Borenstein, Seth. 2025. Why Greenland? Remote but resource-rich island occupies a key position in a warming world, *Associated Press*, https://apnews.com/article/greenland-trump-climate-change-minerals-trade-bab5bb60ba52f6073f056fb271c43215.

Bowen, Thomas et al. 2019. Grid-Scale Battery Storage, *National Renewable Energy Laboratory*, https://www.nrel.gov/docs/fy19osti/74426.pdf.

Boyd, Philip and Vivian, Chris. 2019. Should we fertilize oceans or seed clouds? No one knows, *Nature*, https://www.nature.com/articles/d41586-019-01790-7.

Brant, Robin. 2023. China pledges to stop building new coal energy plants abroad, *BBC News*, https://www.bbc.com/news/world-asia-china-58647481.

Brewster, Murray. 2023. NATO's two per cent spending target — where it came from, what it means, *CBC News*, https://www.cbc.ca/news/politics/nato-military-spending-canada-trudeau-1.6817309.

British Geological Survey. 2023. *Understanding carbon capture and storage*, https://www.bgs.ac.uk/discovering-geology/climate-change/carbon-capture-and-storage/.

British Standards Institution. 2023. *PAS 2060 Carbon Neutrality*, https://www.bsigroup.com/en-GB/PAS-2060-Carbon-Neutrality/.

Brooke, John L. 2014. *Climate Change and the Course of Global History: A Rough Journey*, Cambridge, UK: Cambridge University Press.

Brooks, Chuck. 2023. 3 Alarming Threats To The U.S. Energy Grid – Cyber, Physical, And Existential Events, *Forbes*, https://www.forbes.com/sites/chuckbrooks/2023/02/15/3-alarming-threats-to-the-us-energy-grid--cyber-physical-and-existential-events/?sh=67f4f176101a.

Brown, Benjamin and Samuel J. Spiegel. 2019. Coal, Climate Justice, and the Cultural Politics of Energy Transition, *Global Environmental Politics*, 19(2):149-168, https://direct.mit.edu/glep/article/19/2/149/14931/Coal-Climate-Justice-and-the-Cultural-Politics-of.

Brown, Matthew et al. 2022. Storms batter aging power grid as climate disasters spread, *Associated Press*, https://apnews.com/article/wildfires-storms-science-business-health-7a0fb8c998c1d56759989dda62292379.

Brown University. 2023. *Costs of the 20-year war on terror: $8 trillion and 900,000 deaths*, https://www.brown.edu/news/2021-09-01/costsofwar.

Budinis, Sara. 2022. Direct Air Capture, *International Energy Agency*, https://www.iea.org/reports/direct-air-capture.

Burrows, Leah. 2021. Battery breakthrough for electric cars, *The Harvard Gazette*, https://news.harvard.edu/gazette/story/2021/05/researchers-design-long-lasting-solid-state-lithium-battery/.

Bushard, Brian. 2023. Volkswagen Considers Adopting Tesla's Charging Standards Amid Growing EV Competition—Here Are The Others, *Forbes*, https://www.forbes.com/sites/brianbushard/2023/06/29/polestar-latest-to-adopt-teslas-charging-standards-amid-growing-ev-competition-here-are-the-others/?sh=7ed56caf4afe.

California Air Resources Board. 2023a. *LCFS Data Dashboard*, https://ww2.arb.ca.gov/resources/documents/lcfs-data-dashboard.

California Air Resources Board. 2023b. *Low Carbon Fuel Standard*, https://ww2.arb.ca.gov/our-work/programs/low-carbon-fuel-standard/about.

Calvo, Guiomar et al. 2016. Decreasing Ore Grades in Global Metallic Mining: A Theoretical Issue or a Global Reality?, *Resources* 5(4), https://www.mdpi.com/2079-9276/5/4/36.

Camargo, Luis Ramirez et al. 2022. Pathway to a land-neutral expansion of Brazilian renewable fuel production, *Nature Communications*, 13(3157), https://www.nature.com/articles/s41467-022-30850-2.

Capparella, Joey. 2023. The 25 Bestselling Cars, Trucks, and SUVs of 2023 (So Far), *Car and Driver*, https://www.caranddriver.com/news/g43553191/bestselling-cars-2023/.

CarsDirect. 2023. https://www.carsdirect.com/.

Cavanaugh, Ray. 2017. Peddling purgatory relief: Johann Tetzel, *National Catholic Reporter*, https://www.ncronline.org/news/guest-voices/peddling-purgatory-relief-johann-tetzel.

Central Intelligence Agency. 2023. Country Comparisons – Military Expenditures, *The World Factbook*, https://www.cia.gov/the-world-factbook/field/military-expenditures/country-comparison.

Chang, K. W. and Yoon, H. 2021. Mitigating Injection-Induced Seismicity Along Basement Faults by Extraction: Application to 2016–2018 Pohang Earthquakes, *JGR Solid Earth*, 126(4), e2020JB021486, https://agupubs.onlinelibrary.wiley.com/doi/10.1029/2020JB021486.

Chatterjee, Sudipta et al. 2021. Limitations of Ammonia as a Hydrogen Energy Carrier for the Transportation Sector, *ACS Energy Letters*, 6(12):4390-4394, https://pubs.acs.org/doi/10.1021/acsenergylett.1c02189.

Chiu, Allyson. 2023. What really happens to your clothes after you donate them, *Washington Post*, https://www.washingtonpost.com/climate-solutions/2023/01/04/how-to-donate-clothes-waste-environment/.

Cho, Renee. 2022. With Climate Impacts Growing, Insurance Companies Face Big Challenges, *State of the Planet*, Columbia Climate School, https://news.climate.columbia.edu/2022/11/03/with-climate-impacts-growing-insurance-companies-face-big-challenges/.

Christophers, Brett. 2024. *The Price is Wrong: Why Capitalism Won't Save the Planet*, London: Verso Books.

Chu, Jennifer. 2020. Study: Reflecting sunlight to cool the planet will cause other global changes, *MIT News*, https://news.mit.edu/2020/reflecting-sunlight-cool-planet-storm-0602.

Claes, Julien et al. 2022. Blue carbon: The potential of coastal and oceanic climate action, *McKinsey & Co.*, https://www.mckinsey.com/capabilities/sustainability/our-insights/blue-carbon-the-potential-of-coastal-and-oceanic-climate-action.

Clark, Kevin. 2023. Mitsubishi Power wins order for 1,950 MW of gas-fired power plants in Japan, *Power Engineering*, https://www.power-eng.com/gas/combined-cycle/mitsubishi-power-wins-order-for-1950-mw-of-gas-fired-power-plants-in-japan/.

Clarke, Warren. 2022a. What is a car's out-the-door price?, *Credit Karma*, https://www.creditkarma.com/auto/i/out-the-door-price.

Clarke, Warren. 2022b. What's the average car loan length?, *Credit Karma*, https://www.creditkarma.com/auto/i/car-loan-term.

Clery, Daniel. 2019. Some of the universe's heavier elements are created by neutron star collisions, *Science*, https://www.science.org/content/article/some-universe-s-heavier-elements-are-created-neutron-star-collisions.

Cline, Elizabeth. 2016. The Power of Buying Less by Buying Better, *Atlantic*, https://www.theatlantic.com/business/archive/2016/02/buying-less-by-buying-better/462639/.

Colias, Mike. 2023. GM Scales Back EV Plans as Buyers Hesitate, *Wall Street Journal*, https://www.wsj.com/business/autos/general-motors-gm-q3-earnings-report-2023-5064f4c2.

Congressional Research Service. 2023. *Carbon Storage Requirements in the 45Q Tax Credit*, https://crsreports.congress.gov/product/pdf/IF/IF11639.

Conniff, Richard. 2011. What the Luddites Really Fought Against, *Smithsonian Magazine*, https://www.smithsonianmag.com/history/what-the-luddites-really-fought-against-264412/.

Conway, Ed. 2023. *Material World: The Six Raw Materials that Shape Modern Civilization*, New York: Knopf.

Cooley, Heather and Ajami, Newsha. 2012. Key Issues for Seawater Desalination in California:

Cost and Financing, *Pacific Institute*, https://pacinst.org/wp-content/uploads/sites/21/2013/02/financing_final_report.pdf.

Corbeau, Anne-Sophie and Ann-Kathrin Merz. 2023. Demystifying Electrolyzer Production Costs, *Center on Global Energy Policy*, Columbia University, https://www.energypolicy.columbia.edu/demystifying-electrolyzer-production-costs/.

Coren, Michael. 2023a. Your stove is the first appliance to get a battery, but not the last, *Washington Post*, https://www.washingtonpost.com/climate-environment/2023/03/28/battery-power-home-appliances/.

Coren, Michael. 2023b. Electric vehicles can now power your home for three days, *Washington Post*, https://www.washingtonpost.com/climate-environment/2023/02/07/ev-battery-power-your-home/.

Corporate Finance Institute. 2023. *Carbon Accounting*, https://corporatefinanceinstitute.com/resources/esg/carbon-accounting/.

Council on Environmental Quality. 2016. *Federal Greenhouse Gas Accounting and Reporting Guidance*, https://www.sustainability.gov/pdfs/federal_ghg%20accounting_reporting-guidance.pdf.

Crow, Patrick. 2012. New Report Highlights Staggering Costs Ahead for Water Infrastructure, *WaterWorld*, https://www.waterworld.com/home/article/16193139/new-report-highlights-staggering-costs-ahead-for-water-infrastructure.

Crownhart, Casey. 2025. What's next for nuclear power, *MIT Technology Review*, https://www.technologyreview.com/2025/01/14/1109963/whats-next-for-nuclear-power/.

Crownhart, Casey. 2023. How sodium could change the game for batteries, *MIT Technology Review*, https://www.technologyreview.com/2023/05/11/1072865/how-sodium-could-change-the-game-for-batteries/.

Dagher, Veronica. 2023. What to Do When Your Home-Insurance Policy Isn't Renewed, *Wall Street Journal*, https://www.wsj.com/articles/home-insurance-policy-renewal-6b8a6bf9.

Dalton, Matthew. 2023a. World's First Carbon Import Tax Approved by EU Lawmakers, *Wall Street Journal*, https://www.wsj.com/articles/worlds-first-carbon-tax-approved-by-eu-lawmakers-752ed823.

Dalton, Matthew. 2023b. Switzerland Wants Children to Eat Less Chocolate, More Insects, *Wall Street Journal*, https://www.wsj.com/articles/switzerland-wants-children-to-eat-less-chocolate-more-insects-b1274fa7.

Davis, Jeff. 2017. Infrastructure Week Preview – Federal Infrastructure Grants: How to Get Back to "Average?", *Eno Center for Transportation*, https://www.enotrans.org/article/infrastructure-week-preview-federal-infrastructure-grants-get-back-average/.

Davis, River. 2023. Toyota Chairman Says People Are Finally Seeing the Reality About EVs, *Wall Street Journal*, https://www.wsj.com/business/autos/toyota-chairman-says-people-are-finally-seeing-the-reality-about-evs-31f1669c.

Defense Advanced Research Projects Agency. 2022. *Recovering Rare-Earth Elements from E-Waste*, https://www.darpa.mil/news-events/2022-06-10.

Delbert, Caroline. 2022. This Compact Tokamak Is on the Verge of Commercial Energy Production, *Popular Mechanics*, https://www.popularmechanics.com/science/energy/a39601780/compact-tokamak-fusion-temperature-threshold-commercial-energy/.

Deloitte. 2023. How insurance companies can prepare for risk from climate change, https://www2.deloitte.com/us/en/pages/financial-services/articles/insurance-companies-climate-change-risk.html.

Delmas, Claude. 2018. Sodium and Sodium-Ion Batteries: 50 Years of Research, *Advanced Energy Materials*, 8(17), https://onlinelibrary.wiley.com/doi/abs/10.1002/aenm.201703137.

Dentzer, Bill. 2021. Nevada's Next Boom: High Demand Poised to Spur Silver State's Lithium Production, *Las Vegas Review-Journal*, https://www.reviewjournal.com/local/local-nevada/nevadas-next-boom-demand-poised-to-spur-silver-states-lithium-production-2451259/.

Dezember, Ryan. 2023. Endless Demand Spurs U.S. Natural-Gas Prices to Shale-Era Highs, *Wall Street Journal*, https://www.wsj.com/articles/endless-demand-spurs-u-s-natural-gas-prices-to-shale-era-highs-11661245053.

DiChristopher, Tom. 2023. US spends $81 billion a year to protect global oil supplies, report estimates, *CNBC*, https://www.cnbc.com/2018/09/21/us-spends-81-billion-a-year-to-protect-oil-supplies-report-estimates.html.

DiPippo, Ronald. 2015. *Geothermal Power Plants*, 4th Ed., Oxford, UK: Elsevier Butterworth Heinemann.

D'Odorico, Paolo et al. 2020. The global value of water in agriculture, *Proceedings of the National Academy of Sciences*, 117(36), https://www.pnas.org/doi/10.1073/pnas.2005835117.

Doll, Scooter. 2023a. Here's every electric vehicle that currently qualifies for the US federal tax credit, *Electrek*, https://electrek.co/2023/03/31/which-electric-vehicles-still-qualify-for-us-federal-tax-credit/.

Doll, Scooter. 2023b. "2035 combustion ban moves forward after Germany and EU reach agreement on e-fuels", *Electrek*, https://electrek.co/2023/03/25/2035-combustion-ban-moves-forward-germany-eu-reach-agreement-e-fuels/.

Doll, Scooter. 2022. The top 10 fastest-charging EVs currently available, *Electrek*, https://electrek.co/2022/07/08/fastest-charging-evs/.

Doll, Scooter. 2021. Electric vehicle (EV) charging standards and how they differ, *Electrek*, https://electrek.co/2021/10/22/electric-vehicle-ev-charging-standards-and-how-they-differ/.

Douglas, Erin. 2023. No water, roads or emergency services: How climate change left a rural neighborhood nearly uninhabitable, *Texas Tribune*, https://www.texastribune.org/2023/09/29/texas-climate-managed-retreat-buyouts-liberty-county-flood/.

Driscoll, William. 2023. Electrification could cost California up to $50 billion if loads are not managed, *PV Magazine*, https://pv-magazine-usa.com/2023/05/18/electrification-could-cost-california-up-to-50-billion-if-loads-are-not-managed/.

Duncombe, Jenessa. 2021. What Five Graphs from the U.N. Climate Report Reveal About Our Path to Halting Climate Change, *EOS*, https://eos.org/articles/what-five-graphs-from-the-u-n-climate-report-reveal-about-our-path-to-halting-climate-change.

Duster, Chandelis and Nicole Gaouette. 2021. Biden administration reveals number of nuclear weapons in US stockpile, *CNN*, https://www.cnn.com/2021/10/06/politics/us-nuclear-weapons-stockpile/index.html.

Dvorak, Phred. 2023a. Small Towns Chase America's $3 Trillion Climate Gold Rush, *Wall Street Journal*, https://www.wsj.com/articles/climate-legislation-inflation-reduction-act-small-towns-eb0ce798.

Dvorak, Phred. 2023b. He Pioneered Carbon Offsets to Save Tropical Forests. Now the Market Is Collapsing., *Wall Street Journal*, https://www.wsj.com/science/environment/he-pioneered-carbon-offsets-to-save-tropical-forests-now-the-market-is-collapsing-18a5bc54.

E&E News. 2025. *California carbon offset program under scrutiny*, https://www.eenews.net/articles/california-carbon-offset-program-under-scrutiny/.

Eaglesham, Jean and Pulliam, Susan. 2025. State Farm Was All In on California—Until It Pulled the Plug Before the Fires, *Wall Street Journal*, https://www.wsj.com/business/state-farm-california-pulled-plug-fires-c702fff8.

Eaton, Collin. 2023. Exxon Buys Pipeline Operator, Making Big Bet on Carbon, *Wall Street Journal*, https://www.wsj.com/articles/exxon-buys-pipeline-operator-making-big-bet-on-carbon-a1d0e347.

Eaton, Collin and Morenne, Benoit. 2023. This Arkansas Town Could Become the Epicenter of a U.S. Lithium Boom, *Wall Street Journal*, https://www.wsj.com/articles/this-arkansas-town-could-become-the-epicenter-of-a-u-s-lithium-boom-54ad7306.

Ebert, Roger. 1999. *Reviews: Office Space*, https://www.rogerebert.com/reviews/office-space-1999.

Electricity Advisory Committee. 2019. *Optimizing Reserves: Recommendations for the US Department of Energy*, https://www.energy.gov/sites/default/files/2019/10/f68/EAC_Optimizing%20Reserves%20%28October%202019%29.pdf.

English, Jonathan. 2018. Why Public Transportation Works Better Outside the U.S., *Bloomberg*, https://www.bloomberg.com/news/articles/2018-10-10/why-public-transportation-works-better-outside-the-u-s.

Esqué, Axel et al. 2022. *Decarbonizing the aviation sector: Making net zero aviation possible*, McKinsey & Co., https://www.mckinsey.com/industries/aerospace-and-defense/our-insights/decarbonizing-the-aviation-sector-making-net-zero-aviation-possible.

European Commission. 2023a. *Water Statistics*, https://ec.europa.eu/eurostat/statistics-explained/index.php?title=Water_statistics.

European Commission. 2023b. *EU Emissions Trading System (EU ETS)*, https://climate.ec.europa.eu/eu-action/eu-emissions-trading-system-eu-ets_en.

European Commission. 2023c. *Allocation to industrial installations*, https://climate.ec.europa.eu/eu-action/eu-emissions-trading-system-eu-ets/free-allocation/allocation-industrial-installations_en.

European Commission. 2020. *Start of phase 4 of the EU ETS in 2021: adoption of the cap and start of the auctions*, https://climate.ec.europa.eu/news-your-voice/news/start-phase-4-eu-ets-2021-adoption-cap-and-start-auctions-2020-11-17_en.

European Hydrogen Backbone. 2023. *European Hydrogen Backbone grows to meet REPow-*

erEU's 2030 hydrogen targets, https://www.ehb.eu/newsitem/european-hydrogen-backbone-grows-to-meet-repowereu-s-2030-hydrogen-targets.

European Network for Transmission System Operators for Gas, Gas Infrastructure Europe, and Hydrogen Europe. 2023. *How to Transport and Store Hydrogen – Facts and Figures*, https://www.gie.eu/wp-content/uploads/filr/3429/entsog_gie_he_Qand A_hydrogen_transport_and_storage_210521.pdf.

European Space Agency. 2003. *Greenhouse effects... also on other planets*, https://www.esa.int/Science_Exploration/Space_Science/Venus_Express/Greenhouse_effects_al so_on_other_planets.

Europol. 2009. *Carbon Credit fraud causes more than 5 billion euros damage for European Taxpayer*, https://www.europol.europa.eu/media-press/newsroom/news/carbon-credit-fraud-causes-more-5-billion-euros-damage-for-european-taxpayer.

Evergreen, Shel. 2022. Lithium costs a lot of money—so why aren't we recycling lithium batteries?, *Ars Technica*, https://arstechnica.com/science/2022/04/lithium-costs-a-lot-of-money-so-why-arent-we-recycling-lithium-batteries/.

Ewing, Jack. 2022. Electric Cars Too Costly for Many, Even With Aid in Climate Bill, *New York Times*, https://www.nytimes.com/2022/08/08/business/energy-environment/electric-vehicles-climate-bill.html.

Farràs, Pau, et al. 2021. Water electrolysis: Direct from the sea or not to be?, *Joule*, 5(8):1921-1923, https://www.sciencedirect.com/science/article/pii/S2542435121003524.

Federal Emergency Management Agency. 2023. National Risk Index, https://hazards.fema.gov/nri/map.

Federal Energy Regulatory Commission. 2020. *North American LNG Export Terminals*, https://www.ferc.gov/sites/default/files/2020-11/LNG_Maps_Exports-9-17-2020.pdf.

Findlay, Christopher. 2020. What's your purpose? Reusing gas infrastructure for hydrogen transportation, *Siemens Energy*, https://www.siemens-energy.com/global/en/news/magazine/2020/repurposing-natural-gas-infrastructure-for-hydrogen.html.

Fiorino, Daniel J. 2022. *The Clean Energy Transition: Policies and Politics for a Zero-Carbon World*, Cambridge, UK: Polity Press.

First Street. 2025. *Property Prices in Peril*, https://firststreet.org/research-library/property-prices-in-peril.

Ford, Neil. 2023. EU pledges wind, solar funding to narrow gap with U.S., *Reuters*, https://www.reuters.com/business/energy/eu-pledges-wind-solar-funding-narrow-gap-with-us-2023-01-23/.

Foster, John. 2022. *Realism and the Climate Crisis: Hope for Life*, Bristol, UK: Bristol University Press.

Frangoul, Anmar. 2023. There's a lot of talk about hydrogen's potential. But transportation costs represent a big challenge, *CNBC*, https://www.cnbc.com/2023/07/04/green-hydrogen-is-getting-lots-of-buzz-but-costs-are-a-sticking-point.html.

Freinkel, Susan. 2011. *Plastic: A Toxic Love Story*, New York: Mariner Books.

Freund, Alexander. 2020. Scientists achieve progress on batteries of the future, *Deutsche Welle*, https://www.dw.com/en/the-batteries-of-the-future-sodium-instead-of-lithium/a-54707542.

BIBLIOGRAPHY

Friedmann, Julio et al. 2019. Low-Carbon Heat Solutions for Heavy Industry: Sources, Options, and Costs Today, *Center on Global Energy Policy at Columbia University | SIPA*, https://www.energypolicy.columbia.edu/publications/low-carbon-heat-solutions-heavy-industry-sources-options-and-costs-today/.

Friesen, Kenneth Martens. 2020. *Energy, Economics, and Ethics: The Promise and Peril of a Global Energy Transition*, Lanham, MD: Rowman & Littlefield.

Frueh, Sara. 2021. Infrastructure for a Changing Climate, *National Academies of Science, Engineering and Medicine*, https://www.nationalacademies.org/news/2021/07/infrastructure-for-a-changing-climate.

Fujii-Rajani, Riki and Patnaik, Sanjay. 2025. What will happen to the Inflation Reduction Act under a Republican trifecta?, *Brookings Institution*, https://www.brookings.edu/articles/what-will-happen-to-the-inflation-reduction-act-under-a-republican-trifecta/.

Gaffney, Austyn and Rojanasakul, Mira. 2025. Where Coal Is Retiring, and Hanging On, in the U.S., *New York Times*, https://www.nytimes.com/interactive/2025/02/06/climate/coal-plants-retirement.html.

Gao, Fei-Yue et al. 2022. Seawater electrolysis technologies for green hydrogen production: challenges and opportunities, *Current Opinion in Chemical Engineering*, 36, 100827, https://www.sciencedirect.com/science/article/abs/pii/S2211339822000375.

Gardiner, Beth. 2021. In Europe, a Backlash Is Growing Over Incinerating Garbage, *Yale Environment 360*, https://e360.yale.edu/features/in-europe-a-backlash-is-growing-over-incinerating-garbage.

Garelli, Stéphane. 2023. Why you will probably live longer than most big companies, *International Institute for Management Development*, https://www.imd.org/research-knowledge/disruption/articles/why-you-will-probably-live-longer-than-most-big-companies/.

Gebelhoff, Robert. 2023. A mostly hidden problem wastes appalling amounts of water *Washington Post*, https://www.washingtonpost.com/opinions/2023/03/27/water-utilities-infrastructure-leaks-conservation/.

General Electric. 2023. *Hydrogen fueled gas turbines*, https://www.ge.com/gas-power/future-of-energy/hydrogen-fueled-gas-turbines.

Geoscience Australia. 2023. *Coal*, https://www.ga.gov.au/education/classroom-resources/minerals-energy/australian-energy-facts/coal.

Gershon, Livia. 2021. Gas Shortages in 1970s America Sparked Mayhem and Forever Changed the Nation, *Smithsonian Magazine*, https://www.smithsonianmag.com/smart-news/1970s-gas-shortages-changed-america-180977726/.

Gin Magazine. 2018. *Rate of Distillation*, https://gin-mag.com/2018/09/03/rate-of-distillation/.

Gittleson, Kim. 2012. Can a company live forever?, *BBC News*, https://www.bbc.com/news/business-16611040.

Goodman, A. et al. 2013. Comparison of Publicly Available Methods for Development of Geologic Storage Estimates for Carbon Dioxide in Saline Formations, *NETL-TRS-1-2013*, NETL Technical Report Series, U.S. Department of Energy, National Energy Technology Laboratory, https://netl.doe.gov/projects/files/ComparisonofPublicly AvailableMethodsforDevelopmentofGeologicStorage_031213.pdf.

Götz, Philipp, et al. 2014. Negative Electricity Prices: Causes and Effects, *Energy Brainpool GmbH & Co. KG*, https://www.agora-energiewende.de/fileadmin/Projekte/2013/Agora_Negative_Electricity_Prices_Web.pdf.

Gowdy, John. 2020. Our hunter-gatherer future: Climate change, agriculture and uncivilization, *Futures*, 115, 102488, https://www.sciencedirect.com/science/article/pii/S0016328719303507.

Greenhouse Gas Protocol. 2023. https://ghgprotocol.org/standards.

Groom, Nichola. 2022. U.S. solar expansion stalled by rural land-use protests, *Reuters*, https://www.reuters.com/world/us/us-solar-expansion-stalled-by-rural-land-use-protests-2022-04-07/.

Ground Water Protection Council and Interstate Oil and Gas Compact Commission. 2021. *Potential Induced Seismicity Guide: A Resource of Technical and Regulatory Considerations Associated with Fluid Injection*, https://www.gwpc.org/wp-content/uploads/2022/12/FINAL_Induced_Seismicity_2021_Guide_33021.pdf.

Guardian. 2022. *Extreme UK weather: flights halted as runways melt and temperatures exceed 38C in parts of England – as it happened*, https://www.theguardian.com/uk-news/live/2022/jul/18/uk-weather-heatwave-met-office-warning-forecast-temperature-london-schools-latest-updates.

Guiné, Raquel P. F. et al. 2024. Edible Insects: Consumption, Perceptions, Culture and Tradition Among Adult Citizens from 14 Countries, *Foods*, 13(21), 3408, https://www.mdpi.com/2304-8158/13/21/3408.

Gunia, Amy. 2022. Australia Just Exported Its First Batch of Fuel That Doesn't Emit CO2. There's Just One Catch, *Time*, https://time.com/6139820/liquid-hdyrogen-shipment-hesc/.

Hall, Siobhan. 2023. Hydrogen retrofits for EU gas system cheaper than newbuild: GIE, *S&P Global Commodity Insights*, https://www.spglobal.com/commodityinsights/en/market-insights/latest-news/electric-power/051320-hydrogen-retrofits-for-eu-gas-system-cheaper-than-newbuild-gie.

Hamdy, Louise B. et al. 2021. The application of amine-based materials for carbon capture and utilisation: an overarching view, *Materials Advances*, 2:5843-5880, https://pubs.rsc.org/en/content/articlehtml/2021/ma/d1ma00360g.

Han, Yang and Ho, W. S. Winston. 2021. Polymeric membranes for CO2 separation and capture, *Journal of Membrane Science*, 628, 119244, https://www.sciencedirect.com/science/article/abs/pii/S0376738821001940.

Hanssen, S. V. et al. 2020. The climate change mitigation potential of bioenergy with carbon capture and storage, *Nature Climate Change*, 10:1023-1029, https://www.nature.com/articles/s41558-020-0885-y.

Harford, Tim. 2023. How air conditioning changed the world, *BBC News*, https://www.bbc.com/news/business-39735802.

Harvey, Hal. 2018. *Designing Climate Solutions: A Policy Guide for Low-Carbon Energy*, Washington: Island Press.

Harvey, Hal and Justin Gillis. 2022. *The Big Fix: 7 Practical Steps to Save Our Planet*, New York: Simon & Schuster.

Havlin, Michael. 2023. Automotive dealerships 2019–22: dealer markup increases drive new-vehicle consumer inflation, *Monthly Labor Review*, U.S. Bureau of Labor Statistics, https://doi.org/10.21916/mlr.2023.7.

Heid, Bernd et al. 2022. Five charts on hydrogen's role in a net-zero future, *McKinsey & Co.*, https://www.mckinsey.com/capabilities/sustainability/our-insights/five-charts-on-hydrogens-role-in-a-net-zero-future.

Heinberg, Richard and David Fridley. 2016. *Our Renewable Future: Laying the Path for 100% Clean Energy*, Washington: Island Press.

Heitmann, John A. 2023. Burton Introduces Thermal Cracking for Refining Petroleum, https://ecommons.udayton.edu/cgi/viewcontent.cgi?article=1093&context=hst_fac_pub.

Helm, Dieter. 2020. *Net Zero: How We Stop Causing Climate Change*, London: William Collins.

Henderson, Bob. 2023. California's Green-Fuel Program Gets Too Popular for Its Own Good, *Wall Street Journal*, https://www.wsj.com/articles/californias-green-fuel-program-gets-too-popular-for-its-own-good-dedfb89.

Hennings, P. H. et al. 2021. Pore Pressure Threshold and Fault Slip Potential for Induced Earthquakes in the Dallas-Fort Worth Area of North Central Texas, *Geophysical Research Letters*, 48(15), e2021GL093564, https://agupubs.onlinelibrary.wiley.com/doi/10.1029/2021GL093564.

Herring, David. 2020. Doesn't carbon dioxide in the atmosphere come from natural sources?, *National Oceanic and Atmospheric Administration*, https://www.climate.gov/news-features/climate-qa/doesnt-carbon-dioxide-atmosphere-come-natural-sources.

Hickel, Jason. 2019. Degrowth: a theory of radical abundance, *Real-World Economics Review*, 87:54-68, http://www.paecon.net/PAEReview/issue87/whole87.pdf.

Hill, Alex et al. 2018. How Winning Organizations Last 100 Years, *Harvard Business Review*, https://hbr.org/2018/09/how-winning-organizations-last-100-years.

Hiller, Jennifer. 2025a. Five Things to Know About AI's Thirst for Energy, *Wall Street Journal*, https://www.wsj.com/tech/ai/ai-energy-electricity-use-what-to-know-8c9e64b7.

Hiller, Jennifer. 2025b. Trump Administration Halts Funding for Build-out of Highway EV Chargers, *Wall Street Journal*, https://www.wsj.com/business/autos/trump-administration-halts-funding-for-build-out-of-highway-ev-chargers-da78b791.

Hiller, Jennifer et al. 2024. Tesla Is Pulling Back From EV Charging, and People Are Freaking Out, *Wall Street Journal*, https://www.wsj.com/business/autos/tesla-is-pulling-back-from-ev-charging-and-people-are-freaking-out-ee8e490c.

Hiller, Jennifer et al. 2023. The West Needs Russia to Power Its Nuclear Comeback, *Wall Street Journal*, https://www.wsj.com/articles/nuclear-power-makes-a-comeback-underpinned-by-russian-uranium-24ed8e12?reflink=integratedwebview_share.

Hinsdale, Jeremy. 2022. Cryptocurrency's Dirty Secret: Energy Consumption, *Columbia Climate School*, https://news.climate.columbia.edu/2022/05/04/cryptocurrency-energy/.

Hoefs, Jeremy. 2023. Measurements for an Olympic Size Swimming Pool, *Livestrong*, https://www.livestrong.com/article/350103-measurements-for-an-olympic-size-swimming-pool/.

Hoffmann, Christian et al. 2020. Decarbonization Challenge for Steel, *McKinsey & Co.*, https://www.mckinsey.com/industries/metals-and-mining/our-insights/decarbonization-challenge-for-steel.

Hoffmann, Peter. 2012. *Tomorrow's Energy: Hydrogen, Fuel Cells, and the Prospects for a Cleaner Planet*, Cambridge, MA: MIT Press.

Holger, Dieter. 2023. Coca-Cola Trials Turning Hard-to-Recycle Plastic Into Bottles, *Wall Street Journal*, https://www.wsj.com/articles/coca-cola-trials-turning-hard-to-recycle-plastic-into-bottles-2f8d0dec.

Hood, Abby Lee. 2021. What Made the Battle of Blair Mountain the Largest Labor Uprising in American History, *Smithsonian Magazine*, https://www.smithsonianmag.com/history/battle-blair-mountain-largest-labor-uprising-american-history-180978520/.

Hook, Leslie and Harry Dempsey. 2021. Quest for "green" cement draws big name investors to $300B industry, *Ars Technica*, https://arstechnica.com/science/2021/07/quest-for-green-cement-draws-big-name-investors-to-300b-industry/.

Hoyle, Rhiannon and Julie Steinberg. 2023. Rise of EVs Drives Mining Deals to Decade High, *Wall Street Journal*, https://www.wsj.com/articles/rise-of-evs-drives-mining-deals-to-decade-high-6fab9b47.

Huber, Elena et al. 2024. A qualitative meta-analysis of carbon offset quality criteria, *Journal of Environmental Management*, 352(14), 119983, https://www.sciencedirect.com/science/article/abs/pii/S0301479723027718.

Hughes, Trevor. 2023. The dull way Americans are being forced to care about climate change risk: Insurance, *USA Today*, https://www.usatoday.com/story/news/nation/2023/06/11/climate-change-effects-hit-us-homeowner-insurance/70288893007/.

Hydraulic Institute et al. 2001. Pump Life Cycle Costs: A Guide to LCC Analysis for Pumping Systems, *DOE/GO-102001-1190*, https://www.energy.gov/eere/amo/articles/pump-life-cycle-costs-guide-lcc-analysis-pumping-systems-executive-summary.

Ibrahim, Sara. 2022. Eating insects is still a lot to ask, *SWI*, https://www.swissinfo.ch/eng/society/eating-insects-is-still-a-lot-to-ask/47838316.

Intergovernmental Panel on Climate Change. 2022. *Climate Change 2022: Mitigation of Climate Change—Summary for Policymakers*, https://www.ipcc.ch/report/ar6/wg3/.

International Air Transport Association. 2025. *Voluntary Carbon Offsetting*, https://www.iata.org/en/programs/sustainability/carbon-offset/.

International Energy Agency. 2024. *Bioenergy with Carbon Capture and Storage*, https://www.iea.org/energy-system/carbon-capture-utilisation-and-storage/bioenergy-with-carbon-capture-and-storage.

International Energy Agency. 2023. *Skills Development and Inclusivity for Clean Energy Transitions*, https://www.iea.org/reports/skills-development-and-inclusivity-for-clean-energy-transitions.

International Energy Agency. 2022a. *The Role of Critical Minerals in Clean Energy Transitions, World Energy Outlook Special Report*, https://www.iea.org/reports/the-role-of-critical-minerals-in-clean-energy-transitions.

International Energy Agency. 2022b. *Global data centre energy demand by data centre type, 2010-2022*, https://www.iea.org/data-and-statistics/charts/global-data-centre-energy-demand-by-data-centre-type-2010-2022.

International Energy Agency. 2022c. *Smart Grids*, https://www.iea.org/reports/smart-grids.

International Energy Agency. 2022d. *Global Hydrogen Review 2022*, https://www.iea.org/reports/global-hydrogen-review-2022.

International Energy Agency. 2021a. *Global population without access to electricity by region, 2000-2021*, https://www.iea.org/data-and-statistics/charts/global-population-without-access-to-electricity-by-region-2000-2021-2.

International Energy Agency. 2021b. *End-of-life recycling rates for selected metals*, https://www.iea.org/data-and-statistics/charts/end-of-life-recycling-rates-for-selected-metals.

International Monetary Fund. 2025. *Climate Change: Fossil Fuel Subsidies*, https://www.imf.org/en/Topics/climate-change/energy-subsidies.

International Renewable Energy Agency. 2022. *World Energy Transitions Outlook 2022: 1.5 C Pathway*, https://www.irena.org/publications/2022/Mar/World-Energy-Transitions-Outlook-2022.

Internet Movie Database. 2023. *I Love Lucy – Job Switching*, https://www.imdb.com/title/tt0609243/.

Ip, Greg. 2023. Why No One Wants to Pay for the Green Transition, *Wall Street Journal*, https://www.wsj.com/business/autos/why-no-one-wants-to-pay-for-the-green-transition-aed6ba74.

Jacoby, Tamar. 2014. Why Germany Is So Much Better at Training Its Workers, *Atlantic*, https://www.theatlantic.com/business/archive/2014/10/why-germany-is-so-much-better-at-training-its-workers/381550/.

Jankovska, Olivera and Cohn, Julie A. 2020. Texas CREZ Lines: How Stakeholders Shape Major Energy Infrastructure Projects, *Baker Institute for Public Policy*, https://www.bakerinstitute.org/research/texas-crez-lines-how-stakeholders-shape-major-energy-infrastructure-projects.

Jargon, Julie. 2025. Kids Turn to a Mental-Health Chatbot to Share Their Anxieties, *Wall Street Journal*, https://www.wsj.com/health/wellness/kids-mental-health-chatbot-troodi-d5c646bb.

Jiang, Taoli et al. 2024. Rechargeable Hydrogen Gas Batteries: Fundamentals, Principles, Materials, and Applications, Advanced Materials, 37(1), 2412108, https://advanced.onlinelibrary.wiley.com/doi/10.1002/adma.202412108.

Johnson, Peter. 2025. Honda, Hyundai, Ford, Subaru, and Kia EV sales climb in January, Electrek, https://electrek.co/2025/02/04/honda-hyundai-ford-kia-ev-sales-climb-higher-january/.

Johnson Controls. 2023. *2023 Department of Energy Regulatory Requirements*, https://www.johnsoncontrols.com/doe.

Jordan, Rob. 2021. How does climate change affect migration? *Stanford Earth Matters Magazine*, https://earth.stanford.edu/news/how-does-climate-change-affect-migration.

Joselow, Maxine. 2023. Switching to wind and solar energy will require a lot of land, *Washington Post*, https://www.washingtonpost.com/politics/2023/04/24/switching-wind-solar-energy-will-require-lot-land/.

Joselow, Maxine and Vanessa Montalbano. 2023. Republicans grill Interior chief on mining of critical minerals, *Washington Post*, https://www.washingtonpost.com/politics/2023/03/29/republicans-grill-interior-chief-mining-critical-minerals/.

Joyce, Stephanie. 2014. The Strange History Of The American Pipeline, *KUNC News*, https://www.kunc.org/business/2014-08-05/the-strange-history-of-the-american-pipeline.

JPMorgan Chase. 2013. *J.P. Morgan Ventures Energy Corporation Reaches Settlement with Federal Energy Regulatory Commission,* https://www.jpmorganchase.com/ir/news/2013/id-781214.

Justice, Ginny. 2011. The Role of Indulgences in the Building of New Saint Peter's Basilica, *M.L.S. Thesis, Rollins College,* https://scholarship.rollins.edu/mls/7/.

Kallis, Giorgos et al. 2020. *The Case for Degrowth,* Cambridge, UK: Polity Press.

Kamiya, George. 2022. Data Centres and Data Transmission Networks, *International Energy Agency,* https://www.iea.org/energy-system/buildings/data-centres-and-data-transmission-networks.

Kann, Drew and Nilsen, Ella. 2021. 25% of all critical infrastructure in the US is at risk of failure due to flooding, new report finds, *CNN,* https://www.cnn.com/2021/10/11/weather/infrastructure-flood-risk-climate-first-street/index.html.

Kantchev, Georgi. 2025. What Trump Wants With Greenland, *Wall Street Journal,* https://www.wsj.com/world/europe/what-trump-wants-with-greenland-71ccb535.

Kaplan, Sarah. 2023. Climate change caused catastrophic East Africa drought, scientists say, *Washington Post,* https://www.washingtonpost.com/climate-environment/2023/04/27/climate-change-drought-east-africa/.

Kart, Jeff. 2021. Trees To Capture Carbon (And Cash) In Nation's First Project On State Forest Land, *Forbes,* https://www.forbes.com/sites/jeffkart/2021/09/12/trees-to-capture-carbon-and-cash-in-nations-first-project-on-state-forest-land/?sh=6965db1b53fb.

Kawasaki Heavy Industries. 2023. Hydrogen gas turbine offers promise of clean electricity, *Nature Portfolio,* https://www.nature.com/articles/d42473-022-00211-0.

Keith, David et al. 2020. Reflections on a Meeting About Space-Based Solar Geoengineering, *Solar Geoengineering Research Blog,* Harvard University, https://geoengineering.environment.harvard.edu/blog/reflections-meeting-about-space-based-solar-geoengineering.

Khan, Yusuf. 2023a. Copper Shortage Threatens Green Transition, *Wall Street Journal,* https://www.wsj.com/articles/copper-shortage-threatens-green-transition-620df1e5.

Khan, Yusuf. 2023b. Shipping Giant Maersk Drops Deep Sea Mining Investment, *Wall Street Journal,* https://www.wsj.com/articles/shipping-giant-maersk-drops-deep-sea-mining-investment-c226df39.

Khan, Yusuf and Mackrael, Kim. 2025. Europe Is Looking to Roll Back Climate Accounting Rules, *Wall Street Journal,* https://www.wsj.com/articles/europe-is-looking-to-roll-back-climate-accounting-rules-32c72856.

Kim, June. 2023. Increasing Power Outages Don't Hit Everyone Equally, *Scientific American,* https://www.scientificamerican.com/article/increasing-power-outages-dont-hit-everyone-equally1/.

King, Gilbert. 2012a. Sophie Blanchard – The High Flying Frenchwoman Who Revealed the Thrill and Danger of Ballooning, *Smithsonian Magazine,* https://www.smithsonianmag.com/history/sophie-blanchard-the-high-flying-frenchwoman-who-revealed-the-thrill-and-danger-of-ballooning-89106237/.

King, Gilbert. 2012b. Where the Buffalo No Longer Roamed, *Smithsonian Magazine,* https://www.smithsonianmag.com/history/where-the-buffalo-no-longer-roamed-3067904/.

Kirk, Karin. 2022. Energy loss is single biggest component of today's electricity system,

Yale Climate Connections, https://yaleclimateconnections.org/2022/10/energy-loss-is-single-biggest-component-of-todays-electricity-system/.

Kiss, Daniel et al. 2025. See How Russia Is Winning the Race to Dominate the Arctic, *Wall Street Journal*, https://www.wsj.com/politics/national-security/russia-china-artic-sea-nato-2ca1ea10.

Kleinman, Zoe and Chris Vallance. 2023. Warning AI industry could use as much energy as the Netherlands, *BBC News*, https://www.bbc.com/news/technology-67053139.

Koebele, Elizabeth, and Max Robinson. 2021. The impact of Colorado River water shortages on Nevada, *Nevada Today*, https://www.unr.edu/nevada-today/blogs/2021/impact-of-colorado-river-shortages-on-nevada.

Krok, Andrew. 2022. Toyota Built a Corolla Cross That Uses Hydrogen Like Gasoline, *CNET*, https://www.cnet.com/roadshow/news/toyota-corolla-cross-hydrogen-combustion-concept/.

Larson, Aaron. 2018. Benefits of High-Voltage Direct Current Transmission Systems, *Power Magazine*, August 1, https://www.powermag.com/benefits-of-high-voltage-direct-current-transmission-systems/.

Lauer, Matthew. 2019. The future of work requires a return to apprenticeships, *World Economic Forum*, https://www.weforum.org/agenda/2019/12/apprenticeships-future-work-4ir-training-reskilling/.

Lawrence Berkeley National Laboratory. 2022. *Grid connection requests grow by 40% in 2022 as clean energy surges, despite backlogs and uncertainty*, https://energy.lbl.gov/news/grid-connection-requests-grow-40-2022.

Lawrence Berkeley National Laboratory. 2023. *Queued Up: Characteristics of Power Plants Seeking Transmission Interconnection*, https://emp.lbl.gov/queues.

Layton, Bradley. 2008. A Comparison of Energy Densities of Prevalent Energy Sources in Units of Joules per Cubic Meter, *International Journal of Green Energy*, 5:438-455, https://doi.org/10.1080/15435070802498036.

Lebling, Katie et al. 2022. 6 Things to Know About Direct Air Capture, *World Resources Institute*, https://www.wri.org/insights/direct-air-capture-resource-considerations-and-costs-carbon-removal.

Lee, Jinjoo. 2023. Berkshire Hathaway's Energy Idea Is a Bad Fit for Texas, *Wall Street Journal*, https://www.wsj.com/articles/berkshire-hathaways-energy-idea-is-a-bad-fit-for-texas-c21655a3.

Leswing, Kif. 2019. Apple is spending more than ever on R&D to fulfill the 'Tim Cook doctrine', *CNBC*, https://www.cnbc.com/2019/08/03/apple-rd-spend-increases-fulfilling-tim-cook-doctrine.html.

Lewis, Michelle. 2024. China's first large-scale sodium-ion battery charges to 90% in 12 minutes, *Electrek*, https://electrek.co/2024/05/17/china-first-large-scale-sodium-ion-battery/.

Lewis, Michelle. 2023. Here's what the US needs to do right now to upgrade the grid, *Electrek*, https://electrek.co/2023/05/15/us-grid-upgrade/.

Library of Congress. 2023. *The American West, 1865-1900*, https://www.loc.gov/classroom-materials/united-states-history-primary-source-timeline/rise-of-industrial-america-1876-1900/american-west-1865-1900/.

Lieber, Chavie. 2018. Why fashion brands destroy billions' worth of their own merchandise every year, *Vox*, https://www.vox.com/the-goods/2018/9/17/17852294/fash ion-brands-burning-merchandise-burberry-nike-h-and-m.

Linde, plc. 2019. *Air Separation Plants: History and Technological Progress In the Course of Time*, https://assets.linde.com/-/media/global/engineering/engineering/home/ products-and-services/process-plants/air-separation-plants/air-separation-plants-history-and-technological-progress-2019.pdf.

Lockwood, Deirdre. 2016. Mining landfills for resources doesn't always benefit climate, *Chemical and Engineering News*, https://cen.acs.org/articles/94/web/2016/06/ Mining-landfills-resources-doesnt-always.html.

London Metals Exchange. 2025. *LME Copper*, https://www.lme.com/Metals/Non-ferrous/LME-Copper#Summary.

Macreadie, Peter I. et al. 2021. Blue carbon as a natural climate solution, *Nature Reviews Earth & Environment*, 2:826-839, https://www.nature.com/articles/s43017-021-00224-1.

MacKay, David J. C. 2009. *Sustainable Energy Without the Hot Air*, UIT Cambridge, Ltd., https://www.withouthotair.com/.

MacMillan, Douglas and Will Englund. 2021. Longer, more frequent outages afflict the U.S. power grid as states fail to prepare for climate change, *Washington Post*, https:// www.washingtonpost.com/business/2021/10/24/climate-change-power-outages/.

Maleki, Fariba et al. 2022. Recycling of brass melting slag through the high-temperature oxidation-leaching process, *Sustainable Environmental Research*, 32:2822, https:// sustainenvironres.biomedcentral.com/articles/10.1186/s42834-022-00135-w.

Manwell, J. F. et al. 2009. *Wind Energy Explained: Theory, Design and Application*, New York: John Wiley & Sons.

Marin, Daniela A. et al. 2023. Hydrogen production with seawater-resilient bipolar membrane electrolyzers, *Joule*, 7(4):765-781, https://www.sciencedirect.com/science/ article/abs/pii/S2542435123001174.

Marine Insight. 2024. World's First Liquid Hydrogen-Fueled Oil Tanker Design Gets ClassNK AiP, https://www.marineinsight.com/shipping-news/worlds-first-liquid-hydrogen-fueled-oil-tanker-design-gets-classnk-aip/.

Maritime Executive. 2022. *First International Shipment of Liquid Hydrogen Arrives in Japan*, https://maritime-executive.com/article/first-international-shipment-of-liquid-hydro gen-arrives-in-japan.

24/7 Wall St.. 2020. *How Many Gas Stations Are in U.S.? How Many Will There Be In 10 Years?*, https://247wallst.com/cars-and-drivers/2020/02/16/how-many-gas-stations-are-in-u-s-how-many-will-there-be-in-10-years/.

Martin, Neil. 2022. New system retrofits diesel engines to run on 90 per cent hydrogen, *University of New South Wales at Sydney*, https://www.unsw.edu.au/news/2022/10/ new-system-retrofits-diesel-engines-to-run-on-90-per-cent-hydrog.

Marx, Karl, and Frederick Engels. 2010. *Collected Works: Volume 33*. Lawrence & Wishart Electric Book.

Mayerhöfer, Britta et al. 2020. Bipolar Membrane Electrode Assemblies for Water Electrolysis, *ACS Applied Energy Materials*, 3(10):9635-9644, https://pubs.acs.org/doi/10. 1021/acsaem.0c01127.

Mazzucato, Mariana. 2018. *The Value of Everything: Making and Taking in the Global Economy*, New York: PublicAffairs.

McAleer, Brendan. 2022. Electric Car Battery Life: Everything You Need to Know, *Car and Driver*, https://www.caranddriver.com/research/a31875141/electric-car-battery-life/.

McAllister, Sean. 2024. There could be 1.2 billion climate refugees by 2050. Here's what you need to know, *Zurich Insurance Group*, https://www.zurich.com/media/magazine/2022/there-could-be-1-2-billion-climate-refugees-by-2050-here-s-what-you-need-to-know.

McBride, James et al. 2023. China's Massive Belt and Road Initiative, *Council on Foreign Relations*, https://www.cfr.org/backgrounder/chinas-massive-belt-and-road-initiative.

McKechnie, William Sharp. 2021. Magna Carta: A Commentary on the Great Charter of King John, *Project Gutenberg*, https://www.gutenberg.org/cache/epub/65363/pg65363-images.html.

McKenna, John. 2017. Fancy a bug burger? A Swiss supermarket is selling food made from insects, *World Economic Forum*, https://www.weforum.org/stories/2017/08/fancy-a-bug-burger-a-swiss-supermarket-is-selling-food-made-from-insects/.

McKinlay, Charles J. et al. 2021. Route to zero emission shipping: Hydrogen, ammonia or methanol?, *International Journal of Hydrogen Energy*, 46(55):28282-28297, https://www.sciencedirect.com/science/article/abs/pii/S0360319921022175.

McKinsey & Co. 2023a. *The Inflation Reduction Act: Here's what's in it*, https://www.mckinsey.com/industries/public-and-social-sector/our-insights/the-inflation-reduction-act-heres-whats-in-it.

McKinsey & Co. 2023b. *Renewable-energy development in a net-zero world: Overcoming talent gaps*, https://www.mckinsey.com/industries/electric-power-and-natural-gas/our-insights/renewable-energy-development-in-a-net-zero-world-overcoming-talent-gaps.

McLaughlin, Tim. 2023. Creaky U.S. power grid threatens progress on renewables, EVs, *Reuters*, https://www.reuters.com/investigates/special-report/usa-renewables-electric-grid/.

McNeill, J. R. and Engelke, Peter. 2016. *The Great Acceleration: An Environmental History of the Anthropocene since 1945*, Cambridge, MA: The Belknap Press of Harvard University Press.

McQueen, Noah et al. 2021. A review of direct air capture (DAC): scaling up commercial technologies and innovating for the future, *Progress in Energy*, 3(3), 032001, https://iopscience.iop.org/article/10.1088/2516-1083/abf1ce.

McVeigh, Karen. 2021. Blue carbon: the hidden CO2 sink that pioneers say could save the planet, *Guardian*, https://www.theguardian.com/environment/2021/nov/04/can-blue-carbon-make-offsetting-work-these-pioneers-think-so.

Meichtry, Stacy et al. 2024. Households Wince at the Rising Price of Going Green, *Wall Street Journal*, https://www.wsj.com/science/environment/green-energy-taxes-governments-consumers-7439400d.

Meredith, Sam. 2022. An energy transition loophole is allowing Big Oil to offload high-polluting assets to private buyers, *CNBC*, https://www.cnbc.com/2022/05/19/climate-how-big-oil-sells-off-polluting-assets-in-a-bid-to-look-green.html.

Michals, Debra. 2015. Amelia Earhart, *National Women's History Museum*, https://www.womenshistory.org/education-resources/biographies/amelia-earhart.

Middle East North Africa Financial Network, Inc.. 2022. *How Toyota's Prius Got the Lead Out of Batteries*, https://menafn.com/1105326821/How-Toyotas-Prius-Got-The-Lead-Out-Of-Batteries.

Miettinen, Dylan. 2021. The climate crisis is here. Are insurance companies keeping up?, *Marketplace*, https://www.marketplace.org/2021/08/06/the-climate-crisis-is-here-are-insurance-companies-keeping-up/.

Miller, Richard G. 1996. Estimating global oil resources and their duration, *Norwegian Petroleum Society Special Publications*, 6:43-56, https://www.sciencedirect.com/science/article/abs/pii/S0928893707800051.

Mills, Andrew D. et al. 2019. Impact of Wind, Solar, and Other Factors on Wholesale Power Prices: An Historical Analysis—2008 through 2017, *Lawrence Berkeley National Laboratory*, https://emp.lbl.gov/publications/impact-wind-solar-and-other-factors.

Monnay, Tatyana. 2021. A program that pays farmers not to farm isn't saving the planet, *Politico*, https://www.politico.com/news/2021/08/29/usda-farmers-conservation-program-507028.

Montgomery, Scott L. and Graham, Jr., Thomas. 2017. *Seeing the Light: The Case for Nuclear Power in the 21st Century*, Cambridge, UK: Cambridge University Press.

Mooallem, Jon. 2013. American Hippopotamus, *Atavist Magazine*, 32, https://magazine.atavist.com/american-hippopotamus/.

Morris, Jim. 2025. Justin Trudeau reportedly says Trump's talk of making Canada a US state is 'a real thing', *Associated Press*, https://apnews.com/article/canada-us-tariffs-trudeau-trump-5cfe160c66f90b973d90ea7810cddbe5.

Morton, Oliver. 2015. *The Planet Remade: How Geoengineering Could Change the World*, Princeton, NJ: Princeton University Press.

Moseman, Andrew. 2021. How can carbon emissions from freight be reduced?, *Ask MIT Climate*, https://climate.mit.edu/ask-mit/how-can-carbon-emissions-freight-be-reduced.

Moss, Trefor. 2023. IKEA Redesigns Its Bestsellers, Starting With the Billy Bookcase, *Wall Street Journal*, https://www.wsj.com/articles/ikea-furniture-inflation-billy-c2f835bc.

Mufson, Steven. 2023. Biden's push to disclose climate risks hits wall of industry resistance, *Washington Post*, https://www.washingtonpost.com/climate-environment/2023/05/02/biden-corporate-climate-change-disclosure/.

Najmabadi, Shannon. 2023. Lawmakers Crack Down on Wind-Turbine Lights That Flash All Night, *Wall Street Journal*, https://www.wsj.com/articles/lawmakers-crack-down-on-wind-turbine-lights-that-flash-all-night-21d4b815.

National Academies of Sciences, Engineering, and Medicine. 2023. *Merits and Viability of Different Nuclear Fuel Cycles and Technology Options and the Waste Aspects of Advanced Nuclear Reactors*, Washington, DC: The National Academies Press, https://nap.nationalacademies.org/catalog/26500/merits-and-viability-of-different-nuclear-fuel-cycles-and-technology-options-and-the-waste-aspects-of-advanced-nuclear-reactors.

National Academies of Sciences, Engineering, and Medicine. 2021. *Reflecting Sunlight: Recommendations for Solar Geoengineering Research and Research Governance*, Washington, DC: The National Academies Press, https://nap.nationalacademies.org/catalog/25762/reflecting-sunlight-recommendations-for-solar-geoengineering-research-and-research-governance.

National Aeronautics and Space Administration. 2023a. *Stars,* https://universe.nasa.gov/stars/basics/.

National Aeronautics and Space Administration. 2023b. *Our Solar System,* https://solarsystem.nasa.gov/solar-system/our-solar-system/in-depth/.

National Aeronautics and Space Administration. 2020a. *Milankovitch (Orbital) Cycles and Their Role in Earth's Climate,* https://science.nasa.gov/science-research/earth-science/milankovitch-orbital-cycles-and-their-role-in-earths-climate/.

National Aeronautics and Space Administration. 2020b. *New Evidence Our Neighborhood in Space Is Stuffed With Hydrogen,* https://www.nasa.gov/feature/goddard/2020/new-evidence-our-neighborhood-in-space-is-stuffed-with-hydrogen.

National Aeronautics and Space Administration. 2011. *The Carbon Cycle,* https://www.earthobservatory.nasa.gov/features/CarbonCycle.

National Archives. 2023. *Marshall Plan (1948),* https://www.archives.gov/milestone-documents/marshall-plan.

National Fire Protection Association. 2023. *National Electrical Code,* 406.4(D)(2) Non–Grounding-Type Receptacles, https://www.nfpa.org/codes-and-standards/all-codes-and-standards/list-of-codes-and-standards/detail?code=70.

National Geographic. 2023a. *Should you buy carbon offsets for your air travel?,* https://www.nationalgeographic.com/travel/article/should-you-buy-carbon-offsets-for-your-air-travel.

National Geographic. 2023b. *Wind,* https://education.nationalgeographic.org/resource/wind/.

National Museum of American History. 2023. *Building Ships for Victory,* https://americanhistory.si.edu/on-the-water/answering-call/building-ships-victory.

National Museums Scotland. 2023. *The Newcomen engine and its role in Britain's industrial revolution,* https://www.nms.ac.uk/discover-catalogue/the-newcomen-engine-and-its-role-in-britains-industrial-revolution.

National Oceanic and Atmospheric Administration, 2023. *What is Blue Carbon?,* https://oceanservice.noaa.gov/facts/bluecarbon.html.

National Oceanic and Atmospheric Administration. 2022. *2022 Sea Level Rise Technical Report,* https://oceanservice.noaa.gov/hazards/sealevelrise/sealevelrise-tech-report.html.

National Oceanic and Atmospheric Administration. 2019. *Carbon Cycle,* https://www.noaa.gov/education/resource-collections/climate/carbon-cycle.

National Park Service. 2025. *Producing Plutonium.* https://www.nps.gov/mapr/learn/plutonium.htm.

National Park Service. 2023a. *World War II Shipbuilding in the San Francisco Bay Area,* https://www.nps.gov/articles/000/world-war-ii-shipbuilding-in-the-san-francisco-bay-area.htm.

National Park Service. 2023b. *The Civilian Conservation Corps,* https://www.nps.gov/articles/the-civilian-conservation-corps.htm.

National Renewable Energy Laboratory. 2024. *What Is Power System Curtailment?,* NREL/FS-6A40-90517, https://www.nrel.gov/docs/fy25osti/90517.pdf.

National Renewable Energy Laboratory. 2023a. *Autonomous Energy Systems,* https://www.nrel.gov/grid/autonomous-energy.html.

National Renewable Energy Laboratory. 2023b. *Sensing and Predictive Analytics*, https://www.nrel.gov/grid/sensing-predictive-analytics.html.

National Renewable Energy Laboratory. 2021a. *Beyond Technical Potential: NREL Explores the Challenges of Siting Wind in a Low-Carbon Future*, https://www.nrel.gov/news/program/2021/beyond-technical-potential-nrel-explores-the-challenges-of-siting-wind-in-a-low-carbon-future.html.

National Renewable Energy Laboratory. 2021b. The Challenge of the Last Few Percent: Quantifying the Costs and Emissions Benefits of a 100% Renewable U.S. Electricity System, https://www.nrel.gov/news/program/2021/the-challenge-of-the-last-few-percent-quantifying-the-costs-and-emissions-benefits-of-100-renewables.html.

National Renewable Energy Laboratory. 2020. *A Decade of Transformation: What We Have Learned Since RE Futures Showed What Was Possible*, https://www.nrel.gov/news/features/2022/re-futures.html.

National WWII Museum. 2023. *Ration Books*, https://www.nationalww2museum.org/students-teachers/student-resources/research-starters/take-closer-look-ration-books.

Natter, Ari. 2022. US Seeks $4.3 Billion for Uranium to Wean Off Russia Supply, *Bloomberg*, https://www.bloomberg.com/news/articles/2022-06-07/us-seeks-4-3-billion-for-nuclear-fuel-to-wean-off-russia-supply.

Nature. 2021a. *Concrete needs to lose its colossal carbon footprint*, https://www.nature.com/articles/d41586-021-02612-5.

Nature. 2021b. Give research into solar geoengineering a chance, https://www.nature.com/articles/d41586-021-01243-0.

Nature Climate Change. 2023. Pathways towards 90% decarbonization of aviation by 2050, *Research Briefing*, 12:895-896, https://www.nature.com/articles/s41558-022-01486-3.

Nebergall, Jim. 2022. Hydrogen Internal Combustion Engines and Hydrogen Fuel Cells, *Cummins, Inc.*, https://www.cummins.com/news/2022/01/27/hydrogen-internal-combustion-engines-and-hydrogen-fuel-cells.

Nemetz, Peter. 2022. *Unsustainable World: Are We Losing the Battle to Save Our Planet?*, New York: Routledge.

Neuman, Scott et al. 2023. The fight over the debt ceiling could sink the economy. This is how we got here, *NPR*, https://www.npr.org/2023/03/23/1163448930/what-is-the-debt-ceiling-explanation.

Niermann, M. et al. 2019. Liquid organic hydrogen carriers (LOHCs) – techno-economic analysis of LOHCs in a defined process chain, *Energy & Environmental Science*, 12(1):290-307, https://pubs.rsc.org/en/content/articlelanding/2019/ee/c8ee02700e.

Normile, Dennis. 2021. Environmentalists hail China's vow to stop building coal-fired power plants abroad, *Science*, https://www.science.org/content/article/environmentalists-hail-china-s-vow-stop-building-coal-fired-power-plants-abroad.

Northwest Power and Conservation Council. 2024. *Planning Failure Catalyzes Change*, https://www.nwcouncil.org/regional-power-planning-pacific-northwest/.

Nuclear Energy Agency and Organisation for Economic Cooperation and Development. 2020. Uranium 2020: Resources, Production and Demand, *NEA No. 7551*, https://oecd-nea.org/upload/docs/application/pdf/2020-12/7555_uranium_-_resources_production_and_demand_2020__web.pdf.

Núñez-López, Vanessa and Meckel, Tip. 2021. *Rethinking Geologic Carbon Storage Capacity:*

an approach with practical considerations, presented at the 15th International Conference on Greenhouse Gas Control Technologies, GHGT-15, https://papers.ssrn.com/sol3/papers.cfm?abstract_id=3820949.

Obayashi, Yuka. 2021. Hooked on coal for power, Japan aims for ammonia fix, *Reuters*, https://www.reuters.com/business/energy/hooked-coal-power-japan-aims-ammonia-fix-2021-10-29/.

O'Donovan, Caroline. 2023. Amazon, despite climate pledge, fought to kill emissions bill in Oregon, *Washington Post*, https://www.washingtonpost.com/technology/2023/04/04/amazon-climate-energy-fuel-oregon/.

Oener, Sebastian Z. et al. 2020. Accelerating water dissociation in bipolar membranes and for electrocatalysis, *Science*, 369(6507):1099-1103, https://www.science.org/doi/full/10.1126/science.aaz1487.

Ogg, Clay. 2007. Environmental Challenges Associated With Corn Ethanol Production, US Environmental Protection Agency, *NCEE Working Paper # 07-05*, https://www.epa.gov/sites/default/files/2014-12/documents/environmental_challenges_associated_with_corn_ethanol_production.pdf.

Olah, George et al. 2018. *Beyond Oil and Gas: The Methanol Economy*, Third Edition, New York: Wiley.

O'Neil, Connor. 2022. Small Colorado Utility Sets National Renewable Electricity Example Using NREL Algorithms, *National Renewable Energy Laboratory*, https://www.nrel.gov/news/features/2019/small-colorado-utility-sets-national-renewable-electricity-example-using-nrel-algorithms.html.

Oregon Secretary of State. 2023. *Minimum Energy Efficiency Standards for State-Regulated Appliances and Equipment*, Department of Energy Chapter 330, Division 92, https://secure.sos.state.or.us/oard/viewSingleRule.action?ruleVrsnRsn=296960.

Orf, Darren. 2023. Does the World Have Enough Lithium for Batteries?, *Popular Mechanics*, https://www.popularmechanics.com/science/energy/a42417327/lithium-supply-batteries-electric-vehicles/.

Organisation for Economic Co-operation and Development. 2023. *Plastic pollution is growing relentlessly as waste management and recycling fall short, says OECD*, https://www.oecd.org/environment/plastic-pollution-is-growing-relentlessly-as-waste-management-and-recycling-fall-short.htm.

Osaka, Shannon. 2023a. Here's the biggest hurdle facing America's EV revolution, *Washington Post*, https://www.washingtonpost.com/climate-solutions/2023/04/13/electric-vehicle-charging-network-biden/.

Osaka, Shannon. 2023b. How energy from Earth's crust could pull carbon from the sky, *Washington Post*, https://www.washingtonpost.com/politics/2023/02/23/how-energy-earths-crust-could-pull-carbon-sky/.

Osaka, Shannon. 2023c. A new front in the water wars: Your Internet use, *Washington Post*, https://www.washingtonpost.com/climate-environment/2023/04/25/data-centers-drought-water-use/.

Osaka, Shannon. 2023d. Why the U.S. is so bad at building clean energy, in 3 charts, *Washington Post*, https://www.washingtonpost.com/climate-environment/2023/05/18/permitting-reform-clean-energy-debt-ceiling/.

Osaka, Shannon. 2023e. How this company plans to use Earth's heat to cool the planet,

Washington Post, https://www.washingtonpost.com/climate-solutions/2023/02/23/geothermal-direct-air-capture-fervo/.

Osborne, Margaret. 2023. A Century Ago, This Water Agreement Changed the West. Now, the Region Is in Crisis, *Smithsonian Magazine*, https://www.smithsonianmag.com/smart-news/a-century-ago-this-water-agreement-changed-the-west-now-the-region-is-in-crisis-180981169/.

O'Toole, Randal. 2014. Five Reasons Not to Raise the Gas Tax, *Cato Institute*, https://www.cato.org/commentary/five-reasons-not-raise-gas-tax.

Paddison, Laura. 2023. Greta Thunberg has joined a protest against wind farms. Here's why., CNN, https://www.cnn.com/2023/03/01/europe/greta-thunberg-wind-farm-norway-sami-climate-intl/index.html.

Palumbi, Anthony R. 2023. *At the Base of the Giant's Throat: The Past and Future of America's Great Dams*, Lincoln, NE: University of Nebraska Press.

Panetta, Alexander. 2023. U.S. report claims Trudeau told NATO Canada will never meet its military spending target, *CBC News*, https://www.cbc.ca/news/world/canadian-forces-nato-washington-post-1.6815616.

Papathanasiou, Demetrios. 2023. These developing countries are leading the way on renewable energy, *World Economic Forum*, https://www.weforum.org/agenda/2022/07/renewables-are-the-key-to-green-secure-affordable-energy/.

Parti, Tarini et al. 2025. U.S. Begins Migrant Flights to Guantanamo Bay, *Wall Street Journal*, https://www.wsj.com/politics/policy/trump-immigration-policy-guatanamo-bay-migrant-flights-9fec8df3.

Partlow, Joshua. 2023a. Biden to spend more than $580 million to fix aging water systems in West, *Washington Post*, https://www.washingtonpost.com/climate-environment/2023/04/05/biden-funding-water-infrastructure/.

Partlow, Joshua. 2023b. Water cuts could save the Colorado River. Farmers are in the crosshairs., *Washington Post*, https://www.washingtonpost.com/climate-environment/2023/04/16/colorado-river-crisis-imperial-valley-california/.

Pearson, Hayleigh et al. 2019. Offshore Infrastructure Reuse Contribution to Decarbonisation, *SPE-195772-MS*, Presented at the SPE Offshore Europe Conference and Exhibition, Aberdeen, UK, https://onepetro.org/SPEOE/proceedings-abstract/19OE/2-19OE/D021S007R004/218972.

Peck, David. 2019. A Historical Perspective of Critical Materials, 1939 to 2006, in *Critical Materials: Underlying Causes and Sustainable Mitigation Strategies*, S. Erik Offerman, Ed., World Scientific Series in Current Energy Issues 5, Singapore: World Scientific.

Phillips, Anna. 2023. Why Texas, a clean energy powerhouse, is about to hit the brakes, *Washington Post*, https://www.washingtonpost.com/climate-environment/2023/05/08/texas-solar-wind-power-legislation/.

Picchi, Aimee. 2020. World's billionaires have more wealth than 4.6 billion people, *CBS News*, https://www.cbsnews.com/news/worlds-billionaires-have-more-wealth-than-4point6-billion-people-oxfam-report-today-2020-01-19/.

Piketty, Thomas. 2014. *Capital in the Twenty-First Century*, Cambridge, MA: Belknap Press.

Pitron, Guillaume. 2020. The Rare Metals War: The Dark Side of Clean Energy and Digital Technologies, Victoria, Australia: Scribe Publications.

Piwonka, T. S. 2002. High Performance Materials for Long-Term Usage, *Materials Transac-*

tions, 43(3):376-378, https://www.jstage.jst.go.jp/article/matertrans/43/3/43_3_376/_pdf.

Pleitgen, Fred. 2023. Disappearing lakes, dead crops and trucked-in water: Drought-stricken Spain is running dry, *CNN*, https://www.cnn.com/2023/05/02/europe/spain-drought-catalonia-heat-wave-climate-intl/index.html.

Plimmer, Gill. 2022. Renewables projects face 10-year wait to connect to electricity grid, *Financial Times*, https://www.ft.com/content/7c674f56-9028-48a3-8cbf-c1c8b10868ba.

Plumer, Brad. 2023. The U.S. Has Billions for Wind and Solar Projects. Good Luck Plugging Them In., *New York Times*, https://www.nytimes.com/2023/02/23/climate/renewable-energy-us-electrical-grid.html.

Podesta, John. 2019. The climate crisis, migration, and refugees., *Brookings Blum Roundtable on Global Poverty*, https://www.brookings.edu/research/the-climate-crisis-migration-and-refugees/.

Poonia, Gitanjali. 2021. How the rise of copper reveals clean energy's dark side, *Guardian*, https://www.theguardian.com/us-news/2021/nov/09/copper-mining-reveals-clean-energy-dark-side.

Poynting, Mark. 2024. World's first year-long breach of key 1.5C warming limit, *BBC*, https://www.bbc.com/news/science-environment-68110310.

Preuster, Patrick et al. 2017. Liquid Organic Hydrogen Carriers (LOHCs): Toward a Hydrogen-free Hydrogen Economy, *Accounts of Chemical Research*, 50(1):74-85, https://pubmed.ncbi.nlm.nih.gov/28004916/.

Public Utility Commission of Texas, 2025. *The Texas Energy Fund*, https://www.puc.texas.gov/industry/electric/business/texas-energy-fund/.

Puko, Timothy. 2023a. This power plant offers a peek into the future, *Washington Post*, https://www.washingtonpost.com/climate-environment/2023/05/01/power-plants-hydrogen-climate-change/.

Puko, Timothy. 2023b. EPA plan would impose drastic cuts on power plant emissions by 2040, *Washington Post*, https://www.washingtonpost.com/climate-environment/2023/04/22/epa-power-plant-emissions-climate/.

Qureshi, Ziyad. 2023. 10 Ways Toyota's Hydrogen Combustion Engine Could Revolutionize Performance, *Top Speed*, https://www.topspeed.com/how-toyotas-hydrogen-combustion-engine-could-revolutionize-performance/.

Rahat, Saiful Haque et al. 2024. Bracing for impact: how shifting precipitation extremes may influence physical climate risks in an uncertain future, *Scientific Reports*, 14, 17398, https://www.nature.com/articles/s41598-024-65618-9.

Ramkumar, Amrith. 2023. Fossil-Fuel Veterans Find Next Act With Green Hydrogen, *Wall Street Journal*, https://www.wsj.com/articles/fossil-fuel-veterans-find-next-act-with-green-hydrogen-1f9919d8.

Rao, Rahul. 2021. Physicists want to create energy like stars do. These two ways are their best shot., *Popular Science*, https://www.popsci.com/science/nuclear-fusion-two-methods/.

Reuters. 2023a. *German pipeline firms say they will 'make hydrogen happen' in 2025*, https://www.reuters.com/business/sustainable-business/german-pipeline-firms-say-they-will-make-hydrogen-happen-2025-2022-12-15/.

Reuters. 2023b. *No agreement yet between Germany and EU over e-fuels - transport minister,* https://www.reuters.com/business/sustainable-business/no-agreement-yet-between-germany-eu-over-e-fuels-transport-minister-2023-03-24/.

Reuters. 2021. *Hydrogen, ammonia can help ensure power security in energy transition – IEA,* https://www.reuters.com/business/energy/hydrogen-ammonia-can-help-ensure-power-security-energy-transition-iea-2021-10-06/.

Reuters. 2020. *Inside the world's biggest water desalination plants,* https://www.reuters.com/article/us-saudi-water-desalination-idUSKBN26Y1HD.

Reuters. 2010. *Hungary sells "used" CO2 permits, sparks EU concern,* https://www.reuters.com/article/us-emissions-hungary-idUSTRE62B3UE20100312.

Richardson, Randi et al. Palisades home survived the fire, only to be split in two by a landslide, *NBC News,* https://www.nbcnews.com/weather/wildfires/california-fire-landslide-destroys-palisades-home-rcna188069.

Romm, Cari. 2014. The World War II Campaign to Bring Organ Meats to the Dinner Table, *Atlantic,* https://www.theatlantic.com/health/archive/2014/09/the-world-war-ii-campaign-to-bring-organ-meats-to-the-dinner-table/380737/.

Romo, Vanessa. 2020. PG&E Pleads Guilty On 2018 California Camp Fire: 'Our Equipment Started That Fire', *NPR,* https://www.npr.org/2020/06/16/879008760/pg-e-pleads-guilty-on-2018-california-camp-fire-our-equipment-started-that-fire.

Rosas, Marco. 2025. Chevron openly tells Central Valley drivers why its gas prices are high, *KSEE News,* https://www.yourcentralvalley.com/digital-enterprise/chevron-calls-out-carbon-emissions/.

Ross, Denice et al. 2022. A White House call for real-time, standardized, and transparent power outage data, *White House,* https://bidenwhitehouse.archives.gov/ostp/news-updates/2022/11/22/a-white-house-call-for-real-time-standardized-and-transparent-power-outage-data/.

Royal Museums Greenwich. 2023. *What was the Grand Tour?,* https://www.rmg.co.uk/stories/topics/what-was-grand-tour.

Royal Society of Chemistry. 2023. *Hydrogen,* https://www.rsc.org/periodic-table/element/1/hydrogen.

Rubin, Richard. 2023. Republicans Effectively Voted to Raise Taxes. They're Fine With That., *Wall Street Journal,* https://www.wsj.com/articles/republicans-effectively-voted-to-raise-taxes-theyre-fine-with-that-fb392995.

Saha, Devashree et al. 2023. To Shift Away from Oil and Gas, Developing Countries Need a 'Just Transition' to Protect Workers and Communities, *World Resources Institute,* https://www.wri.org/insights/just-transition-developing-countries-shift-oil-gas.

San Marchi, Chris, and Somerday, Brian P. 2014. Comparison of Stainless Steels for High-Pressure Hydrogen Service, Proceedings of the ASME 2014 Pressure Vessels and Piping Conference, *Paper PVP2014-28811,* https://asmedigitalcollection.asme.org/PVP/proceedings-abstract/PVP2014/46049/V06BT06A023/284260.

Sanders, Robert. 2023. A big step toward 'green' ammonia and a 'greener' fertilizer, *Berkeley News,* https://news.berkeley.edu/2023/01/11/a-big-step-toward-green-ammonia-and-a-greener-fertilizer/.

Sands, Jason. 2015. Why There's Power In Methanol - Racing's Alternative Fuel, *MotorTrend,* https://www.motortrend.com/news/power-in-methanol-racing-alternative-fuel/.

Satran, Joe. 2023. This Is Where America Gets Almost All Its Winter Lettuce, *HuffPost*, https://www.huffpost.com/entry/yuma-lettuce_n_6796398.

Saur, Genevieve et al. 2022. Next Generation Hydrogen Station Composite Data Products: Retail Stations, National Renewable Energy Laboratory, *NREL/PR-5400-83036*, https://www.nrel.gov/docs/fy22osti/83036.pdf.

Schauenberg, Tim. 2023. Deep-sea mining vital to climate action, deadly to oceans, *Deutsche Welle*, https://www.dw.com/en/underwater-mining-to-extract-lithium-cobalt-threatens-biodiversity/a-65219511.

Schechner, Sam. 2025. OpenAI Is Probing Whether DeepSeek Used Its Models to Train New Chatbot, *Wall Street Journal*, https://www.wsj.com/tech/ai/openai-china-deepseek-chatgpt-probe-ce6b864e.

Schechner, Sam and Fitch, Asa. 2025. France Taps Nuclear Power in Race for AI Supremacy, *Wall Street Journal*, https://www.wsj.com/tech/ai/france-taps-nuclear-power-for-new-ai-training-cluster-a7804107.

Scheyder, Ernest. 2023. Copper industry warns of looming supply gap without more mines, *Reuters*, https://www.reuters.com/markets/commodities/copper-industry-warns-looming-supply-gap-without-more-mines-2023-04-20/.

Schneid, Rebecca. 2025. Breaking Down the 'Insider Trading' Accusations Leveled at Trump After Tariffs U-Turn, *Time*, https://time.com/7276515/explaining-insider-trading-accusations-leveled-at-trump-tariffs-pause/.

Seal, Dean. 2025. Edison Unit Says Its Equipment May Have Been Involved in SoCal Fires, *Wall Street Journal*, https://www.wsj.com/business/edison-unit-says-its-equipment-may-have-been-involved-in-socal-fires-32c16040.

Shadel, J. D. 2023. Airlines want you to buy carbon offsets. Experts say they're a 'scam.', *Washington Post*, https://www.washingtonpost.com/travel/2023/04/17/carbon-offsets-flights-airlines/.

Shaw, Kristin. 2023. Lines at Tesla EV chargers are about to get longer, *Popular Science*, https://www.popsci.com/technology/automaker-tesla-ev-charger/.

Sheetz, Michael. 2017. Technology killing off corporate America: Average life span of companies under 20 years, *CNBC*, https://www.cnbc.com/2017/08/24/technology-killing-off-corporations-average-lifespan-of-company-under-20-years.html.

Sherfinksi, David, 2021. U.S. overhauls flood insurance to meet rising climate change risks, *Reuters*, https://www.reuters.com/legal/litigation/us-overhauls-flood-insurance-meet-rising-climate-change-risks-2021-10-15/.

Sheridan, Allison. 2023. The History of the Kerosene Lamp, *Iowa State University History Museums*, https://www.museums.iastate.edu/virtual/blog/2020/04/24/the-history-of-the-kerosene-lamp.

Shrestha, Eriko and Tianyi Sun. 2023. Rule #1 of deploying hydrogen: Electrify first, *Environmental Defense Fund*, https://blogs.edf.org/energyexchange/2023/01/30/rule-1-of-deploying-hydrogen-electrify-first/.

Shrull, Dale. 2023. Working Underground: Men and women coal miners take pride in their unique profession, *The Glenwood Springs Post Independent*, https://www.postindependent.com/news/working-underground-men-and-women-coal-miners-take-pride-in-their-unique-profession/.

Siegel, Rachel and Jeanne Whalen. 2023. New cars, once part of the American Dream, now out of reach for many, *Washington Post*, https://www.washingtonpost.com/busi ness/2023/05/07/new-car-market-high-interest-rates/.

Smart Water Magazine. 2019. *Reverse osmosis systems in industrial processes*, https://smart watermagazine.com/news/membracon/reverse-osmosis-systems-industrial-processes.

Smil, Vaclav. 2017. *Energy and Civilization: A History*, Cambridge, MA: MIT Press.

Smith, Adam B. 2022. 2021 U.S. billion-dollar weather and climate disasters in historical context, *National Oceanic and Atmospheric Administration*, https://www.climate.gov/news-features/blogs/beyond-data/2021-us-billion-dollar-weather-and-climate-disas ters-historical.

Smith, Matthew Nitch. 2016. The number of cars worldwide is set to double by 2040, *World Economic Forum*, https://www.weforum.org/agenda/2016/04/the-number-of-cars-worldwide-is-set-to-double-by-2040.

Smith, Rachel Holliday and Martinez, Jose. 2021. How Does Congestion Pricing Work? What to Know About the Toll System Taking Manhattan, *The City*, https://www.thecity.nyc/2021/9/15/22674371/how-does-congestion-pricing-work-toll-system-in-manhattan.

Smith, Steven Cole. 2021. Running on Empty: There's a Lot to Like about Hydrogen, If You Can Find It, *Car and Driver*, https://www.caranddriver.com/features/a36003212/hydrogen-mirai-california-shortage/.

Smyth, Jamie. 2023. Carbon capture pipeline nixed after widespread opposition, *Ars Technica*, https://arstechnica.com/tech-policy/2023/10/carbon-capture-pipeline-nixed-after-widespread-opposition/.

Snider, Mike. 2025. Trump mints $31 billion with new official $TRUMP crypto meme coin, *USA Today*, https://www.usatoday.com/story/money/2025/01/18/trump-meme-coin-price-crypto/77802704007/.

Society of Petroleum Engineers. 2023. *CO2 Storage Resources Management System*, https://www.spe.org/en/industry/co2-storage-resources-management-system/.

Sol Systems. 2022. *The Sol Standard: Key Insights into LCFS and Clean Fuels Markets*, https://www.solsystems.com/wp-content/uploads/2022/01/The-Sol-Standard-January-25-2022.pdf.

Sommer, Lauren. 2023. Federal climate forecasts could help prepare for extreme rain. But it's years away, *NPR*, https://www.npr.org/2023/01/13/1148854543/flooding-storms-water-infrastructure-climate.

St. John, Alexa and Nora Naughton. 2023a. Auto execs are coming clean: EVs aren't working, *Business Insider*, https://www.businessinsider.com/auto-executives-coming-clean-evs-arent-working-2023-10.

St. John, Alexa and Nora Naughton. 2023b. EVs are running out of customers — and some dealers don't want them anymore, *Business Insider*, https://www.businessin sider.com/dealers-turning-away-evs-velectric-cars-demand-cools-inventory-2023-8.

Steil, Benn. 2019. *The Marshall Plan: Dawn of the Cold War*, New York: Simon & Schuster.

Stevens, Harry. 2023a. America needs clean electricity. These states show how to do it., *Washington Post*, https://www.washingtonpost.com/climate-environment/interac tive/2023/clean-energy-electricity-sources/.

Stevens, Harry. 2023b. Trees are moving north from global warming. Look up how your city could change., *Washington Post*, https://www.washingtonpost.com/climate-environment/interactive/2023/tree-species-climate-change-north-shift/.

Stevens, Harry. 2023c. We need an area the size of Texas for wind and solar. Here's how to halve it., *Washington Post*, https://www.washingtonpost.com/climate-environment/interactive/2023/renewable-energy-land-use-wind-solar/.

Stowe, Michael L. 2017. Compressed Air Basics, *Chemical Engineering Progress*, https://www.aiche.org/resources/publications/cep/2017/may/compressed-air-basics.

Strickland, Eliza. 2019. With "Leapfrog" Technologies, Africa Aims to Skip the Present and Go Straight to the Future, *IEEE Spectrum*, https://spectrum.ieee.org/with-leapfrog-technologies-africa-aims-to-skip-the-present-and-go-straight-to-the-future.

Strovink, Kurt. 2021. Reimagine Insurance, *McKinsey & Co.*, https://www.mckinsey.com/industries/financial-services/our-insights/how-insurance-can-help-combat-climate-change.

Sullivan, Tim. 2025. What to know about Guantánamo Bay, the base where Trump will send 'criminal aliens', *Associated Press*, https://apnews.com/article/guantanamo-bay-detention-migrants-what-to-know-trump-d027c5c24b523f31a62271dcbe7c010e.

Suneson, Grant. 2020. Big-name brands that disappeared in the last decade include Borders, Pier 1 Imports and Toys R Us, *USA Today*, https://www.usatoday.com/story/money/2020/12/31/brands-that-disappeared-in-the-last-decade/43298485/.

Taggart, Sue. 2023. How Food Has Changed in The Past 50 Years, *Eco18*, https://eco18.com/how-food-has-changed-in-the-past-50-years/.

Tarascon, Jean-Marie. 2020. Na-ion versus Li-ion Batteries: Complementarity Rather than Competitiveness, *Joule*, 4(8), https://www.sciencedirect.com/science/article/pii/S2542435120302403.

Taylor, Michael. 2021. Analysis-Indonesia's palm oil-powered 'green diesel' fuels threat to forests, *Reuters*, https://www.reuters.com/article/us-indonesia-climate-biodiesel-idUSKBN2A4030.

Terlouw, Tom et al. 2022. Large-scale hydrogen production via water electrolysis: a techno-economic and environmental assessment, *Energy & Environmental Science*, 15(9):3583-3602, https://pubs.rsc.org/en/content/articlelanding/2022/ee/d2ee01023b.

Tomer, Adie et al. 2021. America has an infrastructure bill. What happens next?, *Brookings Institution*, https://www.brookings.edu/blog/the-avenue/2021/11/09/america-has-an-infrastructure-bill-what-happens-next/.

Tongia, Rahul. 2022. It is unfair to push poor countries to reach zero carbon emissions too early, *Brookings Institution*, https://www.brookings.edu/blog/planetpolicy/2022/10/26/it-is-unfair-to-push-poor-countries-to-reach-zero-carbon-emissions-too-early/.

Tooze, Adam. 2023. Hydrogen Is the Future—or a Complete Mirage, *Foreign Policy*, https://foreignpolicy.com/2023/07/14/hydrogen-is-the-future-or-a-complete-mirage/.

Tracy, Ben and Novak, Analisa. 2023. Carbon capture technology: The future of clean energy or a costly and misguided distraction?, *CBS News*, https://www.cbsnews.com/news/carbon-capture-climate-change-clean-energy/.

Treviño, R. H., and Meckel, T. A., editors. 2017. Geological CO2 Sequestration Atlas of

Miocene Strata, Offshore Texas State Waters, *Report of Investigations No. 283*, The University of Texas at Austin, Bureau of Economic Geology.

Tweed, Katherine. 2015. Electricity Use Could Soar as Global Middle Class Embraces Air Conditioning, *IEEE Spectrum*, https://spectrum.ieee.org/electricity-consumption-could-soar-as-global-middle-class-embraces-ac.

Union of Concerned Scientists. 2023. *How Natural Gas Is Formed*, https://www.ucsusa.org/resources/how-natural-gas-formed.

Union of Concerned Scientists. 2020. *What Is Solar Geoengineering?*, https://www.ucsusa.org/resources/what-solar-geoengineering.

United Nations. 2021. *Climate and weather related disasters surge five-fold over 50 years, but early warnings save lives - WMO report*, https://news.un.org/en/story/2021/09/1098662.

United Nations Environment Programme. 2019. *Three ways we can better use nitrogen in farming*, https://www.unep.org/news-and-stories/story/three-ways-we-can-better-use-nitrogen-farming.

United Nations Environment Programme. 2020. *Fertilizers: challenges and solutions*, https://www.unep.org/news-and-stories/story/fertilizers-challenges-and-solutions.

United Nations Framework Convention on Climate Change. 2024. *COP29 Agrees International Carbon Market Standards*, https://unfccc.int/news/cop29-agrees-international-carbon-market-standards.

United Nations Framework Convention on Climate Change. 2023. *COP26 Outcomes: Market mechanisms and non-market approaches (Article 6)*, https://unfccc.int/process-and-meetings/the-paris-agreement/the-glasgow-climate-pact/cop26-outcomes-market-mechanisms-and-non-market-approaches-article-6.

United Nations Regional Information Centre for Western Europe. 2025. *Artificial intelligence: How much energy does AI use?*, https://unric.org/en/artificial-intelligence-how-much-energy-does-ai-use/.

University of Central Florida. 2023. *Food Tourism: The impact of food TV shows on local industries*, https://www.ucf.edu/online/hospitality/news/food-tourism/.

University of Michigan. 2023a. *Critical Materials Factsheet, Center for Sustainable Systems*, https://css.umich.edu/publications/factsheets/material-resources/critical-materials-factsheet.

University of Michigan. 2023b. *Photovoltaic Energy Factsheet*, https://css.umich.edu/publications/factsheets/energy/photovoltaic-energy-factsheet.

University of Michigan. 2023c. *Wind Energy Factsheet*, https://css.umich.edu/publications/factsheets/energy/wind-energy-factsheet.

U.S. Bureau of Land Management. 2023. *El Camino Real de Tierra Adentro National Historic Trail*, https://www.blm.gov/visit/el-camino-real-de-tierra-adentro-national-historic-trail.

U.S. Bureau of Reclamation. 2023a. *Colorado River Compact, 1922*, https://www.usbr.gov/lc/region/g1000/pdfiles/crcompct.pdf.

U.S. Bureau of Reclamation. 2023b. *Near-term Colorado River Operations – Draft Supplemental Environmental Impact Statement*, https://www.usbr.gov/ColoradoRiverBasin/SEIS.html.

U.S. Census Bureau. 2023. *Quick Facts*, https://www.census.gov/quickfacts/fact/table/US/VET605221.

BIBLIOGRAPHY

U.S. Census Bureau. 2022. *Median Household Income*, https://www.census.gov/library/visualizations/2022/comm/median-household-income.html.

U.S. Census Bureau. 1998. *Population of the 100 Largest Cities and Other Urban Places In The United States: 1790 to 1990*, https://www.census.gov/library/working-papers/1998/demo/POP-twps0027.html.

U.S. Department of Agriculture. 2023a. *US Food Imports*, https://www.ers.usda.gov/data-products/u-s-food-imports/.

U.S. Department of Agriculture. 2023b. *Croplands in a Changing Climate*, https://www.climatehubs.usda.gov/croplands-changing-climate.

U.S. Department of Defense. 2020. *During WWII, Industries Transitioned From Peacetime to Wartime Production*, https://www.defense.gov/News/Feature-Stories/story/Article/2128446/during-wwii-industries-transitioned-from-peacetime-to-wartime-production/.

U.S. Department of Energy. 2024. *DOE Releases New Report Evaluating Increase in Electricity Demand from Data Centers*, https://www.energy.gov/articles/doe-releases-new-report-evaluating-increase-electricity-demand-data-centers.

U.S. Department of Energy. 2023a. *Air-Source Heat Pumps*, https://www.energy.gov/energysaver/air-source-heat-pumps.

U.S. Department of Energy. 2023b. "Incorporate Minimum Efficiency Requirements for Heating and Cooling Products into Federal Acquisition Documents", https://www.energy.gov/femp/incorporate-minimum-efficiency-requirements-heating-and-cooling-products-federal-acquisition.

U.S. Department of Energy. 2023c. *Safe Use of Hydrogen*, https://www.energy.gov/eere/fuelcells/safe-use-hydrogen.

U.S. Department of Energy. 2023d. *Hydrogen Shot*, https://www.energy.gov/eere/fuelcells/hydrogen-shot.

U.S. Department of Energy. 2023e. *Hydrogen Fueling Station Locations*, https://afdc.energy.gov/fuels/hydrogen_locations.html.

U.S. Department of Energy. 2022. *Grid Energy Storage: Supply Chain Deep Dive Assessment*, https://www.energy.gov/sites/default/files/2022-02/Energy%20Storage%20Supply%20Chain%20Report%20-%20final.pdf.

U.S. Department of Energy. 2009. Hydrogen Fueling—Coming Soon to a Station Near You, *DOE/GO-102009-2812*, https://www.nrel.gov/docs/fy09osti/45407.pdf.

U.S. Department of Homeland Security. 2021. *Feature Article: Securing Transportation of Ammonia—Agricultural Lifeline and Future Affordable, Clean Energy Source*, https://www.dhs.gov/science-and-technology/news/2021/06/17/feature-article-securing-transportation-ammonia.

U.S. Department of Transportation. 2014. *Will Americans Support Fuel Tax Increases? The Answer Could Be Surprising*, https://www.transportation.gov/utc/will-americans-support-fuel-tax-increases-answer-could-be-surprising.

U.S. Department of the Treasury. 2025. What is the national debt?, https://fiscaldata.treasury.gov/americas-finance-guide/national-debt/.

U.S. Energy Information Administration. 2023a. *Oil and petroleum products explained*, https://www.eia.gov/energyexplained/oil-and-petroleum-products/.

U.S. Energy Information Administration. 2023b. *Global Oil Markets*, https://www.eia.gov/outlooks/steo/report/global_oil.php.

U.S. Energy Information Administration. 2023c. *What is the efficiency of different types of power plants?*, https://www.eia.gov/tools/faqs/faq.php?id=107&t=3.

U.S. Energy Information Administration. 2023d. *How much electricity is lost in electricity transmission and distribution in the United States?*, https://www.eia.gov/tools/faqs/faq.php?id=105&t=3.

U.S. Energy Information Administration. 2023e. *Electric Power Monthly*, Data for August 2023, https://www.eia.gov/electricity/monthly/.

U.S. Energy Information Administration. 2023f. *U.S. simple-cycle natural gas turbines operated at record highs in summer 2022*, https://www.eia.gov/todayinenergy/detail.php?id=55680.

U.S. Energy Information Administration. 2021. *U.S. electricity customers experienced eight hours of power interruptions in 2020*, https://www.eia.gov/todayinenergy/detail.php?id=50316.

U.S. Environmental Protection Agency. 2023a. *U.S. Electricity Grid & Markets*, https://www.epa.gov/green-power-markets/us-electricity-grid-markets.

U.S. Environmental Protection Agency. 2023b. *Greenhouse Gases at EPA*, https://www.epa.gov/greeningepa/greenhouse-gases-epa.

U.S. Environmental Protection Agency. 2021. *EPA Report: U.S. Cars Achieve Record High Fuel Economy and Low Emission Levels as Companies Fully Comply with Standards*, https://www.epa.gov/newsreleases/epa-report-us-cars-achieve-record-high-fuel-economy-and-low-emission-levels-companies.

U.S. Geological Survey. 2023a. *How much water does the typical hydraulically fractured well require?*, https://www.usgs.gov/faqs/how-much-water-does-typical-hydraulically-fractured-well-require.

U.S. Geological Survey. 2023b. *Lithium*, https://pubs.usgs.gov/periodicals/mcs2021/mcs2021-lithium.pdf.

U.S. Geological Survey. 2023c. *Rare Earths*, https://pubs.usgs.gov/periodicals/mcs2022/mcs2022-rare-earths.pdf.

U.S. Geological Survey. 2022. *U.S. Geological Survey Releases 2022 List of Critical Minerals*, https://www.usgs.gov/news/national-news-release/us-geological-survey-releases-2022-list-critical-minerals.

U.S. Geological Survey. 2019a. *Critical Mineral Commodities in Renewable Energy*, https://www.usgs.gov/media/images/critical-mineral-commodities-renewable-energy.

U.S. Geological Survey. 2019b. *Mining Water Use*, https://www.usgs.gov/mission-areas/water-resources/science/mining-water-use.

U.S. Geological Survey. 2018. *Where is Earth's Water?*, https://www.usgs.gov/special-topics/water-science-school/science/where-earths-water.

U.S. Geological Survey. 2011. *Wind Energy in the United States and Materials Required for the Land-Based Wind Turbine Industry From 2010 Through 2030*, https://pubs.usgs.gov/sir/2011/5036/sir2011-5036.pdf.

U.S. Securities and Exchange Commission. 2022. *SEC Proposes Rules to Enhance and Standardize Climate-Related Disclosures for Investors*, https://www.sec.gov/news/press-release/2022-46.

Valero, Alicia et al. 2021. *The Material Limits of Energy Transition: Thanatia*, Cham: Springer Nature Switzerland.

Valinsky, Jordan. 2024. Three Mile Island is reopening and selling its power to Microsoft,

CNN, https://www.cnn.com/2024/09/20/energy/three-mile-island-microsoft-ai/index.html.

Van Dam, Andrew. 2023. The real reason trucks have taken over U.S. roadways, *Washington Post*, https://www.washingtonpost.com/business/2023/04/07/trucks-outnumber-cars/.

Van der Kroon, Bianca et al. 2011. *The energy ladder: Theoretical myth or emperical truth?*, IVM Institute for Environmental Studies, VU University Amsterdam.

Van Ness, H. C. 1983. *Understanding Thermodynamics*, New York: Dover Publications, Inc.

Verma, Pranshu. 2023. Even electric self-driving cars may have a climate change problem, *Washington Post*, https://www.washingtonpost.com/technology/2023/01/13/self-driving-cars-greenhouse-gas-emissions/.

Verne, Jules. 1994. *Twenty Thousand Leagues Under the Sea*. Project Gutenberg, https://www.gutenberg.org/ebooks/164.

Vincent, John M. 2023. How Much Do Electric Cars Cost?, *U.S. News & World Report*, https://cars.usnews.com/cars-trucks/advice/electric-car-prices.

Voelcker, John. 2022. Hydrogen Fuel-Cell Vehicles: Everything You Need to Know, *Car and Driver*, https://www.caranddriver.com/features/a41103863/hydrogen-cars-fcev/.

Von Meier, Alexandra. 2006. *Electric Power Systems: A Conceptual Introduction*, New York: Wiley-IEEE Press.

Wagner, H. and Fettweis, G. B. L. 2001. About science and technology in the field of mining in the Western world at the beginning of the new century, *Resources Policy*, 27:157-168, https://www.sciencedirect.com/science/article/abs/pii/S0301420701000162.

Wall Street Journal. 2023. *California's Income Tax on Electricity*, https://www.wsj.com/articles/california-democrats-income-electricity-bill-energy-climate-68aec815.

Wallace, Harold D., Jr. 2023. Power from the people: Rural Electrification brought more than lights, *National Museum of American History*, https://americanhistory.si.edu/blog/rural-electrification.

Wang, Qiang et al. 2022. Impact of COVID-19 pandemic on oil consumption in the United States: A new estimation approach, *Energy*, 239 Part C, 122280, https://www.sciencedirect.com/science/article/pii/S0360544221025287.

Wang, Yang, et al. 2022. A review of low and zero carbon fuel technologies: Achieving ship carbon reduction targets, *Sustainable Energy Technologies and Assessments*, 54, 102762, https://www.sciencedirect.com/science/article/abs/pii/S2213138822008104.

Wang, Zhongmin and Alan Krupnick. 2013. US Shale Gas Development: What Led to the Boom?, *Resources for the Future*, Issue Brief 13-4, https://media.rff.org/archive/files/sharepoint/WorkImages/Download/RFF-IB-13-04.pdf.

France 24. 2023. *Water-starved Saudi confronts desalination's heavy toll*, https://www.france24.com/en/live-news/20230917-water-starved-saudi-confronts-desalination-s-heavy-toll.

Wei, Yi-Ming et al. 2021. A proposed global layout of carbon capture and storage in line with a 2°C climate target, *Nature Climate Change*, 11:112-118, https://www.nature.com/articles/s41558-020-00960-0.

Wesoff, Eric and Jonathan Gifford. 2020. Top global Li-ion battery projects: Tesla grows lead as Hornsdale expands to 150 MW, *PV Magazine*, https://pv-magazine-usa.com/2020/04/18/top-8-global-li-ion-battery-projects-tesla-grows-lead-with-hornsdale-expansion-to-150-mw/.

White House. 2023a. *FACT SHEET: Biden-Harris Administration Announces New Standards and Major Progress for a Made-in-America National Network of Electric Vehicle Chargers*, https://www.presidency.ucsb.edu/documents/fact-sheet-biden-harris-administration-announces-new-standards-and-major-progress-for-made.

White House. 2023b. *Inflation Reduction Act Guidebook*, https://bidenwhitehouse.archives.gov/cleanenergy/inflation-reduction-act-guidebook/.

Wildlife Trusts. 2023. *Peatlands*, https://www.wildlifetrusts.org/natural-solutions-climate-change/peatland.

Wood, Johnny. 2021. Eager To Become Hydrogen-Ready, Power Plants Turn To Dual-Fuel Turbines, *Forbes*, https://www.forbes.com/sites/mitsubishiheavyindustries/2021/07/30/eager-to-become-hydrogen-ready-power-plants-turn-to-dual-fuel-turbines/?sh=24a6eeb34760.

Woods, Darren. 2021. Why we're investing $15 billion in a lower-carbon future, *Exxon-Mobil Corp.*, https://corporate.exxonmobil.com/news/news-releases/2021/1109_why-we-are-investing-15-billion-in-a-lower-carbon-future.

World Bank. 2023a. *Climate-Smart Agriculture*, https://www.worldbank.org/en/topic/climate-smart-agriculture.

World Bank. 2023b. *GDP (current US$)*, https://data.worldbank.org/indicator/NY.GDP.MKTP.CD?year_high_desc=false.

World Bank. 2023c. *GDP per capita (current US$) - United States*, https://data.worldbank.org/indicator/NY.GDP.PCAP.CD?locations=US.

World Bank. 2023d. *Carbon Pricing Dashboard*, https://carbonpricingdashboard.worldbank.org/.

World Bank. 2023e. *Oil rents (% of GDP)*, https://data.worldbank.org/indicator/NY.GDP.PETR.RT.ZS.

World Bank. 2023f. *Coal rents (% of GDP)*, https://data.worldbank.org/indicator/NY.GDP.COAL.RT.ZS.

World Bank. 2023g. *Natural gas rents (% of GDP)*, https://data.worldbank.org/indicator/NY.GDP.NGAS.RT.ZS.

World Economic Forum. 2022. *The 200-year history of mankind's energy transitions*, https://www.weforum.org/agenda/2022/04/visualizing-the-history-of-energy-transitions/.

World Economic Forum. 2017. *Landfill mining: is this the next big thing in recycling?*, https://www.weforum.org/agenda/2017/06/landfill-mining-recycling-eurelco/.

Xing, Hui et al. 2021. Alternative fuel options for low carbon maritime transportation: Pathways to 2050, *Journal of Cleaner Production*, 297, 126651, https://www.sciencedirect.com/science/article/abs/pii/S0959652621008714.

Yahoo! Finance. 2024. Natron Energy Achieves First-Ever Commercial-Scale Production of Sodium-Ion Batteries in the U.S., https://finance.yahoo.com/news/natron-energy-achieves-first-ever-160200575.html.

Yale Climate Connections. 2021. *Maritime shipping causes more greenhouse gases than airlines*, https://yaleclimateconnections.org/2021/08/maritime-shipping-causes-more-greenhouse-gases-than-airlines/.

Yale Law School. 2021. *Human Rights Workshop: Guantanamo Litigator on Lawyering in a*

Lawless Space, https://law.yale.edu/yls-today/news/human-rights-workshop-guan tanamo-litigator-lawyering-lawless-space.

Yergin, Daniel. 1991a. Blood and Oil: Why Japan Attacked Pearl, *Washington Post,* https://www.washingtonpost.com/archive/opinions/1991/12/01/blood-and-oil-why-japan-attacked-pearl/1238a2e3-6055-4d73-817d-baf67d3a9db8/.

Yergin, Daniel. 1991b. *The Prize: The Epic Quest for Oil, Money & Power,* New York: Simon & Schuster.

Yu, Mingquan et al. 2022. Principles of Water Electrolysis and Recent Progress in Cobalt-, Nickel-, and Iron-Based Oxides for the Oxygen Evolution Reaction, *Angewandte Chemie,* 61(1), e202103824, https://onlinelibrary.wiley.com/doi/10.1002/anie.202103824.

Zewe, Adam. 2023. Computers that power self-driving cars could be a huge driver of global carbon emissions, *MIT News,* https://news.mit.edu/2023/autonomous-vehi cles-carbon-emissions-0113.

Zhang, Zong-Xian et al. 2021. World mineral loss and possibility to increase ore recovery ratio in mining production, International Journal of Mining, Reclamation and Environment, 35(9), https://www.tandfonline.com/doi/full/10.1080/17480930.2021.1949878.

ABOUT THE AUTHOR

Thomas Manuel Ortiz is an engineer with 30 years of diverse experience in energy, from hydrogen/solar cogeneration research and the use of recycled carbon dioxide as a net-zero refrigerant to oil and gas exploration, production and refining to offshore wind siting and carbon sequestration. Tom, author of the popular Substack newsletter Resource Realism, holds a Ph.D. in mechanical engineering from Purdue University and an M.B.A. in finance from Texas A&M University. He is also a registered professional engineer in the State of Texas. He lives in Austin with his wife.

Looking for your next book?
We publish the stories you've been waiting to read!

Scan the QR code below to get 20% off your next Stoney Creek title!

For author book signings, speaking engagements, or other events,
please contact us at info@stoneycreekpublishing.com

StoneyCreekPublishing.com